George F. Hammond

A Treatise on Hospital and Asylum Construction

With Special Reference to Pavilion Wards

George F. Hammond

A Treatise on Hospital and Asylum Construction
With Special Reference to Pavilion Wards

ISBN/EAN: 9783337163426

Printed in Europe, USA, Canada, Australia, Japan

Cover: Foto ©Andreas Hilbeck / pixelio.de

More available books at **www.hansebooks.com**

A

TREATISE

ON

HOSPITAL AND ASYLUM CONSTRUCTION;

WITH

SPECIAL REFERENCE TO PAVILION

WARDS.

BY

GEORGE F. HAMMOND,

ARCHITECT.

VOL. I.

SPECIAL PRELIMINARY EDITION.

CLEVELAND, OHIO,

1891.

CONTENTS.

ILLUSTRATIONS CLASSIFIED.

NEW YORK HOSPITAL, FIFTEENTH STREET.

OPERATING ROOMS—Continued.

INEXPENSIVE OPERATING AND EMERGENCY ANNEX.

COMPARISONS OF CIRCULAR AND RECTANGULAR WARDS.

CANCER HOSPITAL, NEW YORK.

WARREN WARD, MASSACHUSETTS GENERAL HOSPITAL, BOSTON.

OCTAGONAL WARD, JOHNS HOPKINS HOSPITAL, BALTIMORE.

RECTANGULAR WARD, JOHNS HOPKINS HOSPITAL, BALTIMORE.

PREFACE.

A friend, to whom a selection from this book was sent before "going to press," wrote, "I do not know whether you submit the quotation from your forthcoming book with intent of having it criticised or not, but I fear that reviews of the book will contain criticisms of what you say." Having prepared it from an interest in the subject and because of the pleasure derived from jotting down one's thoughts on paper, rather than with a premeditated idea of publication, the prospects of its being honored by a "review" had not been considered by the writer. The answer may be quoted from the Autocrat of the Breakfast Table. "I wonder if anybody ever finds fault with anything I say at this table when it is repeated? I hope they do, I am sure. I should be very certain that I had said nothing of much significance, if they did not."

As it is not written with the expectation of making every physician his own architect, the plates are merely outlines intended to visibly demonstrate the letter press; therefore.flues, air spaces and other necessary details are not, as a rule, indicated. The book's "raison d' être" is the fact that while some physicians have acquainted themselves with the subject, more have—very properly—concentrated their thoughts on matters pertaining more directly to their private practice, and who, when they are called on for advice on the subject (as they are liable to be at any time) naturally turn for reference to the earlier works on the subject.

From them they are liable to infer that the fan for heating or ventilating is "a weak invention of the enemy"; that a fire place, instead of being a valuable adjunct on account of its *ventilating* qualities, is the only proper means of *heating* a ward; that because old fashioned plumbing is offensive no modern work can be sanitarily correct, and must be banished to a semi-detached pavilion, which while desirable as affording an additional security, the writer does not consider essential if reasonable attention is paid to the cleanliness of exterior surfaces.

He further believes that while the pavilion hospital should invariably be built when the site will admit, other methods of planning *may* be sanative, *if carefully studied,* for "half a loaf (well prepared) is better than no bread." He further believes that a successful plan is more important than an imposing exterior, though the latter should not be sacrificed if it can be helped.

While Hospitals and Asylums are so closely allied that the dividing line is scarcely discernible, it seemed advisable to separate these subjects in this work, although much relating to one is equally applicable to the other.

This division is all the more difficult to determine because the trend of professional parlance is to apply the term "Hospitals" to all institutions for the treatment of the insane. But while the writer is in accord with this practice, he has adhered to the older and more significant term "Asylums" to which is devoted the second section of this work.

A plea is made for specialism in both hospital and asylum construction, with the belief that soon special hospitals for the treatment of Gynaecological and Lying-In cases and special asylums for the treatment and segregation of Epileptics and also Insane Criminals, will be recognised as advisable.

The writer makes no pretensions to any medical knowledge beyond that acquired in the study of hospital construction, and the few medical terms have been used because they best convey his meaning. Reports or information concerning newly erected or proposed hospitals or asylums will be gratefully acknowledged by him.

Having thus expressed the object of this work, it is sent forth with the hope that it may add its mite to the knowledge of these buildings; the subscriber being solely responsible for the status of the original theories advanced herein.

Cleveland, Ohio, GEORGE F. HAMMOND,
May, 1891. ARCHITECT.

NOTE.

The writer desires to express his obligations to the following physicians for letters of information, or courtesies extended at the numerous hospitals visited by him.

H. H. Powell, M. D., Cleveland.

R. B. Dixon, M. D., Boston.

G. H. M. Rowe, M. D., Supt. and Res. Physician, City Hospital, Boston.

Edward Cowles, M. D., Med. Supt., McLean Asylum, Somerville.

John W. Pratt, M. D., Res. Physcian, Mass. Gen. Hospital, Boston.

George P. Ludlam, M. D., Supt. New York Hospital, New York.

Jas. R. Lathrop, M. D., Supt. Roosevelt Hospital, New York.

Willard D. Becker, M. D., House Surgeon, Cancer Hospital, New York.

Henry M. Hurd, M. D., Supt. Johns Hopkins Hospital, Baltimore.

P. M. Wise, M. D., Med. Supt. St. Lawrence State Hospital, Ogdensburg.

William A. Hammond, M. D., Washington.

John B. Hamilton, U. S. Marine Hospital Service, Washington.

Georges Angenot, M. D., Civil Hospital, Antwerp.

Herbert Jones, B. A., M. B., M. R. C. S., Miller Memorial, Greenwich, Eng.

And to the following architects for letters of information and drawings of the following buildings :

Charles C. Haight, New York, Cancer Hospital.

Rand and Taylor, Boston, Mary Hitchcock Hospital.

W. Wheeler Smith, New York, McLane Operating Room.

Alfred Waterhouse, London, Royal Infirmary, Liverpool.

And to other unknown architects whose buildings are represented herein.

NOTE

SPECIAL PRELIMINARY EDITION.

This volume is one of a small preliminary edition which the writer has had printed for convenience in making references in the succeeding volumes.

For this reason and because it has been prepared during the interim and subject to the interposing duties of Architectural engagements, no careful revision of grammatical and typographical errors has yet been made.

The scope of the entire work will be substantially as follows:

VOL. I. Hospitals. General Introductory.

VOL. II. Asylums. General Introductory.

VOL. III. Construction and Sanitary Devices.

VOL. IV. Special Departments and Stationary Fixtures.

The letter-press and some of the illustrations of the first three volumes are ready and will be published with such additional volumes as the writer's time and means will permit.

GEORGE F. HAMMOND,

Cleveland, O.,   ARCHITECT.

September, 1891.

CHAPTER I.

TYPES OF HOSPITALS.

Hospitals may be divided into three general types according to their outline or plan, viz: First, Block planned; Second, Corridor planned; Third, Pavilion planned. Until a comparatively recent date hospitals were planned on either the Block or Corridor system, the former being the most common. Now it is obsolete so far as planning new buildings is concerned, and it is to be hoped that it will not be many years before the use of the Block hospital is abandoned.

It consisted of any irregular shaped building of large size, say rectangular, subdivided by interior walls into connecting wards : this made it frequently necessary to pass through one ward to get to another, although, of course, some had entrances on halls. It also prevented the wards from having light on both sides, while beds were placed against the interior partitions that were perforated by occasional door openings. These admitted the air from one ward to another, by reason of the draft, to be breathed by the patients of both wards in common.

The two latter systems must not be confounded: although both may have corridors, the corridor system *per se* consists of a building having a corridor along one side or in the center, with wards opening from it, the corridor partition forming one side of the ward. The wards were therefore lighted on one side only, beds were arranged on both, and there was no noticeable difference in the interior appearence of the wards of the block and corridor plans, although the marked difference in separating them by a corridor is easily perceptible.

In the pavilion plan, the corridor still connects with the wards, but they are separated from each other so that light is admitted to them on two and usually three sides, their ends connect with the corridor, and windows are placed in the corridor walls between the wards.

These are the typical characteristics of the three classes of hospitals, although modifications of each have been erected that were great improvements on their prototypes.

Plate 1, page 3 shows the three systems.

Fig. 1, Block plan; Fig. 2, Corridor plan; Fig. 3, Pavilion plan.

The writer has prepared a modification of the Block plan in plate 2, page 5, which resembles in some respects the Pavilion plan. It is possible that it might be of some slight value as an opthalmic

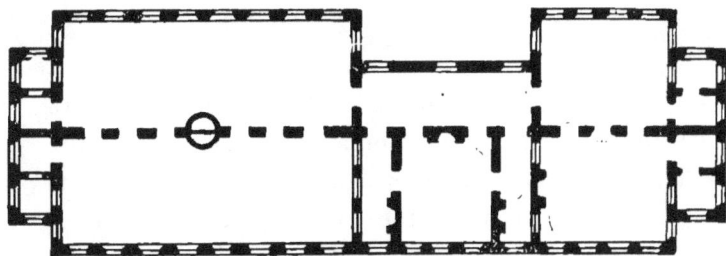

FIG. 1.

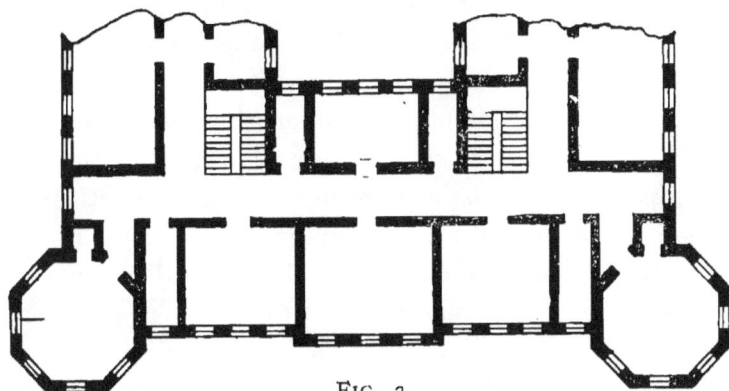

FIG. 2.

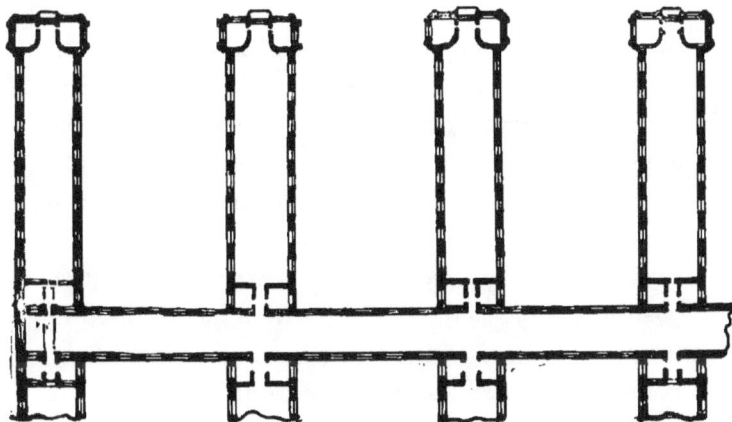

FIG. 3.
PAGE 3.   PLATE 1.

ward, or for convalescent cases of measles or scarlet fever, where strong light is objectionable, as there would be no glare from opposite windows; but the temptation to place beds against the blank wall as shown, and to change the ward to *general* use, makes it objectionable and renders the plan dangerous. It is shown here as an example of what should *not* be done unless beds are confined to the spaces between, or next to windows.

A modification of the corridor plan, resembling the pavilion plan in some respects, is shown in plate 3, page 7, and is now being erected from the writer's plans. It is his expectation, that eventually semi-detached pavilion wards will be erected in connection with it, as shown by dotted lines.

It will be noticed that the larger wards have light on three sides, and the single rooms on each side of the corridor are intended for mild cases: the building is two stories in height, one floor being devoted to males, the other to females. Surgical cases will occupy one side and medical cases the other.

The pavilion plan has been modified in two ways of many stages each: first, as regards the shape of pavilions, second, as regards their relative position. The shape of the pavilion has resulted in the circular ward as the latest form of construction. While considered experimental until recently, its growth has been gradual, but steady.

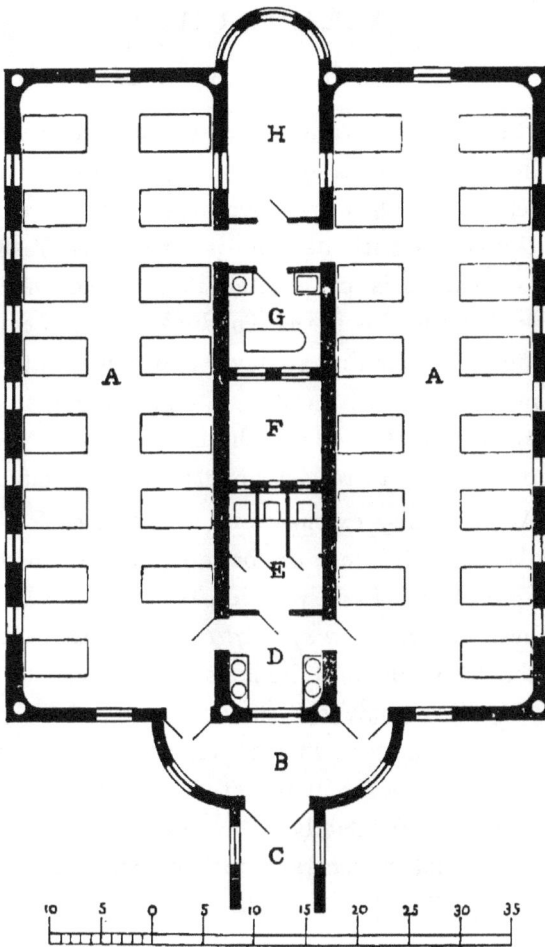

PLATE 2, PAGE 5.

An Improper Arrangement.

A, Ward; B, Lobby; C, Connecting Corridor: D,
Lavatory; G, Bath Room; H, Nurse.

CHAPTER II.

SHAPES OF PAVILION WARDS.

The shape of pavilion—and of course in pavilion hospitals this means the ward,— varies greatly, and each form has its advocates. Of course the pavilion itself with its semi-isolation, its fresh air and sunshine, is of greater importance than its shape; though this too is worthy of serious consideration. All pavilions may be divided into the following classes. First, oblong, in which the length is usually about three times the width; second, square; third, octagonal; fourth, circular.

While many examples of wards of each shape may be found in block or corridor planned hospitals dating earlier than all pavilion construction, it was by these stages, *when applied to pavilions*, that the circular form which is the latest, was reached.

The development of the circular pavilion hospital was slow, and its success was assured only after repeated failures of many other forms,—although the failures were usually on account of their arrangement or relative location rather than because of their shape.

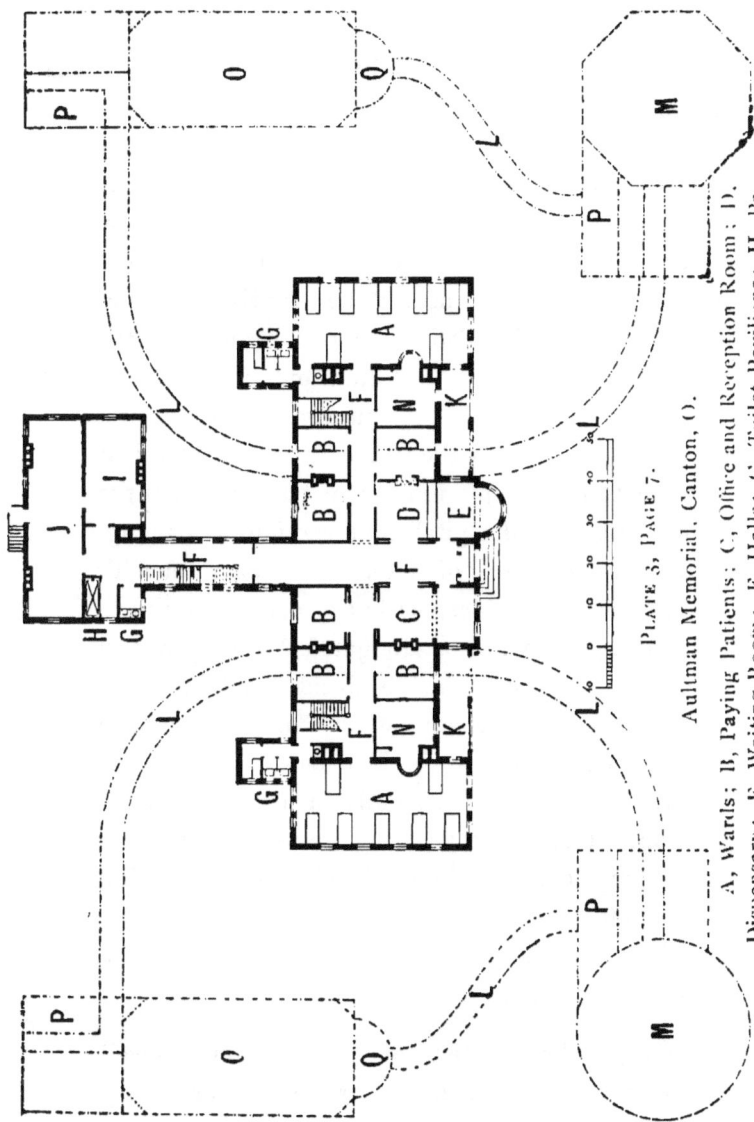

PLATE 3, PAGE 7.

Aultman Memorial. Canton, O.

A, Wards; B, Paying Patients; C, Office and Reception Room; D, Dispensary; E, Waiting Room; F, Halls; G, Toilet Pavilions; H, Patients' Elevator; I, Bakery; J, Kitchen; K, Piazzas; L, Connecting Platforms; M, O, Pavilions; P, Piazzas; Q, Sun Rooms. See also pages 137, 139, 141 and 143.

Among the first buildings attempted with rectangular pavilions, was the often referred to Lariboisiere hospital, built in Paris in 1854 ; it was mostly three stories in height, but the stairs were so arranged that there was direct communication by them from one floor to another so that no isolation of ward air could be obtained.   The new Hotel Dieu in Paris, completed in 1876, is another example of the many-storied pavilion hospital, but it has been severely criticised by French authorities on Hospital work.   It was during the Franco–Prussian war, just previous to the completion of this building, that the Tollet system of hospital construction was inaugurated.   It consisted of a one story pavilion with windows on both sides, with the corners rounded, and with a cylindrical or elongated dome ceiling in the ward.   The roof had a decided pitch, being usually covered with shingles or slate.   Along one side it was customary to build a balcony or piazza. The toilet apparatus was built in a semi-detached pavilion.   The entire building was set on a series of posts or piers so as to admit of a free circulation of air under as well as around it.   It was usually heated by a stove or stoves set in the centre of the ward, or by fire places.

This Tollet system of construction is well adapted for military or field uses, and for temporary rather than for permanent buildings in this country

PLATE 4, PAGE 9.

Perspective of Aultman Memorial.

where the climate is so variable and severe : it was the progenitor of most modern pavilion hospitals of this class.   Plate 5, page 11.

It was also about this time, or shortly previous, that a plan for an octagonal ward was produced by the late John R. Niernsee, an architect of Baltimore. The ceiling of this pavilion ward was arranged to pitch in straight lines towards the centre, and fire-places were located in the centre for ventilation as well as heating.   It contained beds for 24 patients and had a diameter of a little over 60 feet.   The advantage of a ward of this shape can be readily seen, as it admits sunlight from all points of the compass during the entire day if it is properly located.

It has in connection with it certain administra-tion rooms, and they are designed to be connected by a continuous enclosed corridor leading to the other similar pavilions.

While some portions of this plan are open to criticism, it at that time represented a step some-what in advance of those mentioned before.   The diameter of the ward (60 feet or more), seems to be excessive ; as a rule from 30 to 45 feet are sufficient, and it would seem injudicious to construct these wards of a greater diameter, than 50 feet.   Where a larger number of beds *must* be accommodated in one ward than can be accomodated in a ward of this

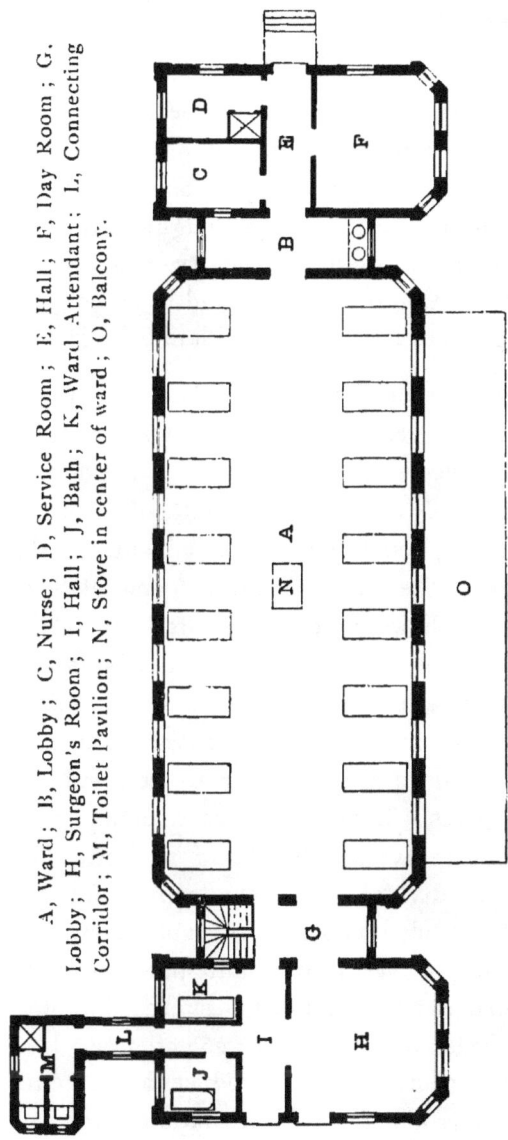

A, Ward; B, Lobby; C, Nurse; D, Service Room; E, Hall; F, Day Room; G, Lobby; H, Surgeon's Room; I, Hall; J, Bath; K, Ward Attendant; L, Connecting Corridor; M, Toilet Pavilion; N, Stove in center of ward; O, Balcony.

PLATE 5, PAGE 11.
Toilet Plan.

diameter, it would be preferable to erect an oblong
shaped ward: because although the beds are placed
at the walls, the centre of the ward should be well
lighted, and twenty-five feet is quite as far as it will
well illuminate in wards of usual height.   Another
reason is that there would be difficulty in obtaining
an equal distribution of fresh air, and in preventing
areas of stagnation in the ward.   Plate 6, page 13.

We will now consider a ward that is in many
characteristics different from any other, viz: a circular
ward with a domed or sloping ceiling.   An example
of it may be found in the Civil Hospital of Antwerp,
plate 12, page 25, which shows to advantage the
skeleton plan of the entire building, and plate 7,
page 15, which shows an enlarged drawing of the
arrangement of one of the wards or pavilions.   It
represented the newest form of pavilion construction.
It attracted much attention because it was the first
building planned in this manner to be used for
hospital purposes.   Each ward has a diameter of
61 ½ feet and an average height of 17 feet.   The
space which is divided off in the centre for the
nurse is not intended as a sleeping apartment.   It
is fitted up with shelves on which are put bottles for
medicines, dishes and other articles needed in the
ward.   The beds are arranged for 26 patients.   It is
claimed that any domed ceiling produces air currents
which materially aid natural ventilation,—particu-

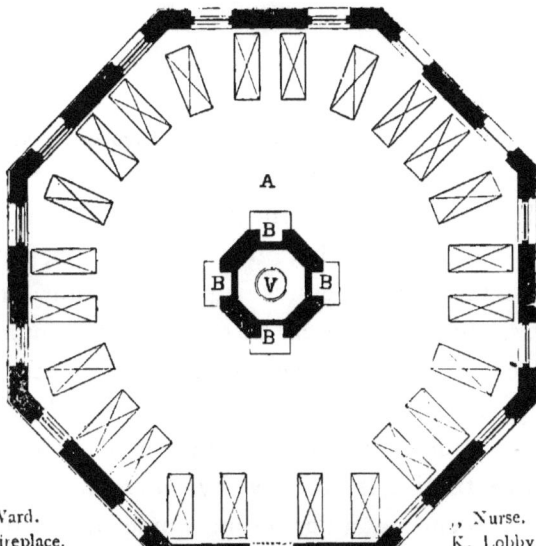

A, Ward.
B, Fireplace.
C, Scullery.
D, Lobby.
E, Water Closets
F, Sink.
G, Linen.
H, Closet.
I, Dumb Waiter.

J, Nurse.
K, Lobby.
L, Bowles,
  General Bath,
  Vapor and
  Shower Bath.
M, Patients'
  Wardrobe.
N, Hall.
O, Connecting
  Corridor.
P, Ventilating
  Shaft.

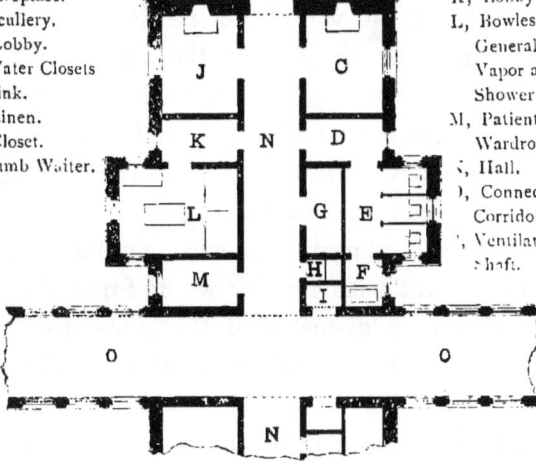

PLATE 6, PAGE 13.
Niernsee's Octagonal Ward.

larly if the dome is not too flat. This was noticed in church buildings used temporarily as hospitals many years ago, but the discovery does not seem to have been utilized until recently.

It is undoubtedly true that these currents of air exist in rooms with ceilings of this form, but to what extent they possess superior sanitary conditions, cannot very well be decided on account of the lack of unity in size, heating and ventilating arrangements of wards of this shape, which would render it difficult to form a true comparison. But if the ventilation is obtained by *mechanical* means, it would seem that a flat ceiling would be sanitarily correct because the ventilation is assured and *continuous*, and does not require the aid of a sloping form to lead to the apertures in the ceiling, in order to facilitate the escape of foul air and gases thereby.

Immediately following the Antwerp hospital was commenced the construction of the Miller Memorial hospital near Greenwich, England : it too, was constructed with circular wards containing ten beds and with a diameter of thirty-five feet. On some accounts it attracted more attention than the Antwerp Civil Hospital, and in a letter from the resident physician in 1887, the writer was informed that the success attending this ward, its conveniences and its facilities for light and air, were of so superior a nature that the erection of another was contemplated in the near future.   Plate 9, page 19.

A, Ward; B, Ward Utensils; c, Heating Columns; C, Ventilating Column; D, Connecting Lobby; E. F. Separation Wards; G, Hall; H, Stair-well; I, Stairs; J, Poultice room containing chute to basement; K, Separation Ward;

L, Elevator; M. Stairs to space above ward and roof; N, Diet Kitchen; O, Bath; P, Lavatory; Q, Bowls; R, Urinals; S, Water Closets; T, Scullery; U, Corridor roof connecting with other wards.

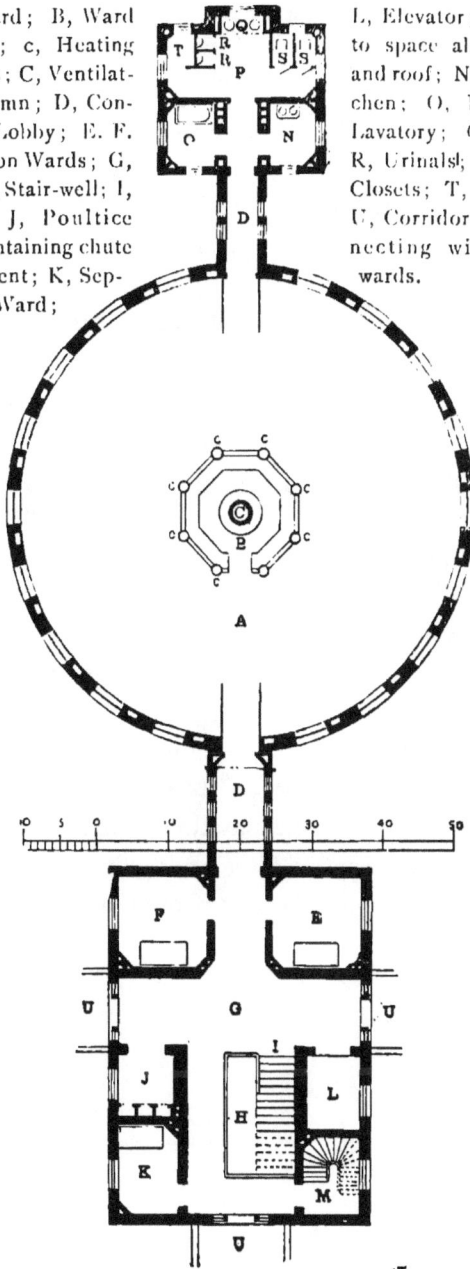

PLATE 7, PAGE 15.

Circular Ward, Civil Hospital, Antwerp.

See also Plate 12, Page 25.

CHAPTER III.

As regards the relative position of pavilions, the most popular is the arrangement in which their medial lines are parallel with each other. The reasons for this are, that in even the largest institutions, sufficient land to place them any other way and secure the requisite accomodations is frequently unobtainable, and because they all bear the same relative position to the points of the compass and receive the sun alike. This arrangement may be advantageous for oblong pavilions receiving the sun principally on one or two sides, but is not of so much importance in locating octagonal or circular pavilions which may receive light and sunshine equally well from any direction.

In all hospitals there have been attempts to completely isolate surgical and medical wards. With the best pavilion plans, there has been, according to the number of stories, a distance of 50 to 100 feet between the pavilions when placed parallel with each other. This is undoubtedly sufficient, because contagious diseases should not be received

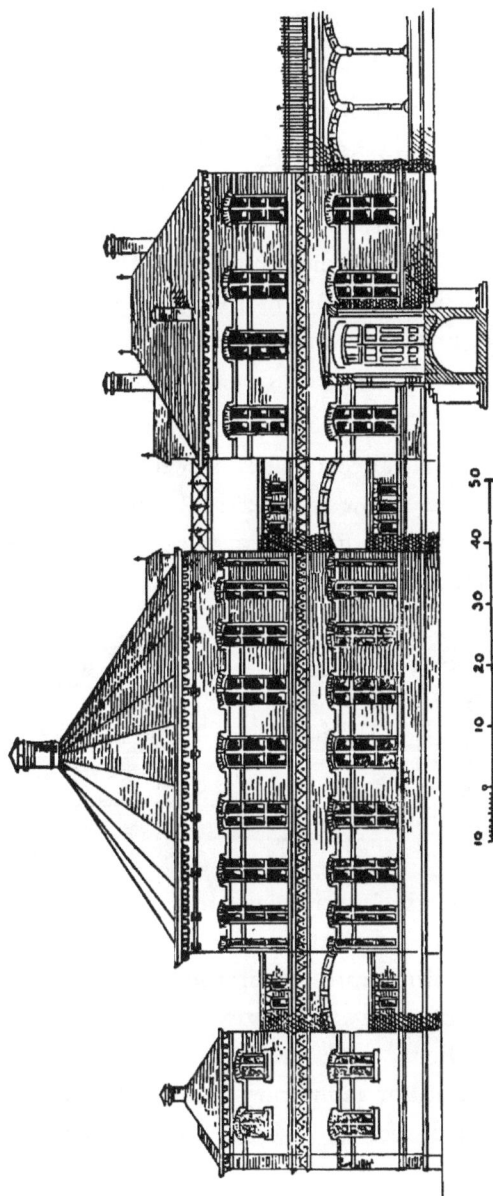

PLATE 8, PAGE 17.

Elevation of One Ward of Civil Hospital, Antwerp.

in common with other cases. Regarding the distance between them, it may be governed entirely by
the height from the first floor to the top of cornice,
and most writers have decided (when they have
mentioned it at all) that twice the height of the
pavilion is sufficient (meaning from the ground).
It should be borne in mind that this is to allow
direct sunlight to shine into the sides of the lower
wards as early and as late as possible, as well as to
secure satisfactory isolation.

Attention is called to the following cut showing
the advantage of a low pitched roof, and the writer
has adopted for his guidance a rule which is given
here for what it may be worth.

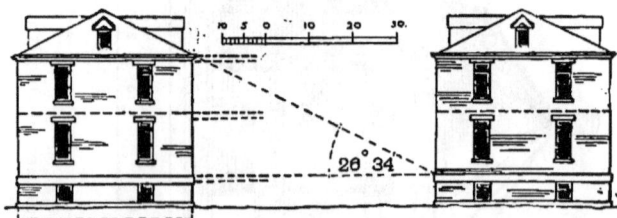

Make the inclination of the roof 26° 34′
(quarter pitch), and continue the line of it until it
intercepts the line representing the lower floor level.
This will give the minimum distance between
parallel pavilions, and prevent the attics being
improperly used for temporary wards. Of course a
steeper pitched roof should not change the *angle*

A, Ward;
B, Fire Places; C, Toilet Pavilion;
D, Lobby;
E, Ward;
Scullery; F,
Separation
Ward: G,
Bath Room;
H, Hall; I,
Closet; J,
Stair-well;
K, Stairs;
L, M, Matron.

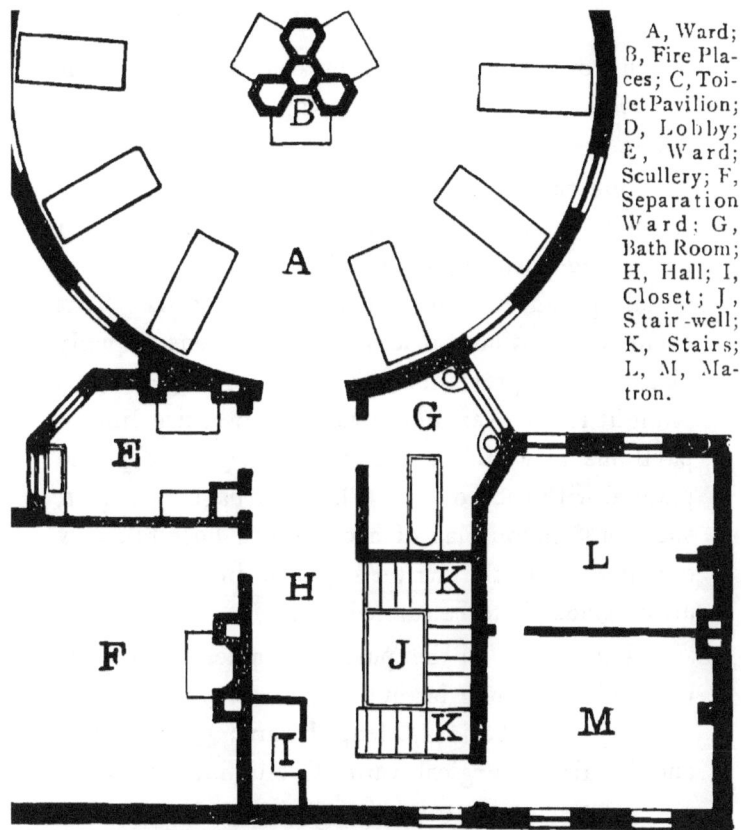

Second Floor Plan, Miller Memorial Hospital,
Greenwich, Eng.

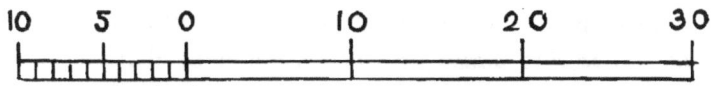

PLATE 9, PAGE 19.

which would then be made by drawing a line from the *ridge* of the roof to intercept the floor line as before.

But the value of octagonal or circular wards seems to the writer to be enhanced by the curved or half circle position, with radiating and circumferential platforms, not enclosed by walls. This ensures isolation, places the pavilions equally distant, and prevents the wind from getting a straight range over and through the entire line of pavilions, as would be the case if they were placed parallel with each other. Plate 10, page 21, fig. 1, shows a skeleton plan of a children's hospital, now erecting from the writer's plans, in which this arrangement is to be carried out.

The figures on the pavilions indicate the uses to which they may be put. 1, Measles; 2, Scarlet fever; 3, Diphtheria;4, Medical ward; 5, Opthalmic and Aural; 6, Surgical ward; C, Administration.

As the buildings are being erected from time to time, modifications of these divisions may be made and the writer considers the administration building—as planned—too small, *though as extensive as the funds will admit;* he has recommended its enlargement as in modified plan on plate 10, fig. 2.

The first octagonal pavilion on this plan erected by the writer, is the one built in connection with the Lakeside Hospital in Cleveland, and is a temporary

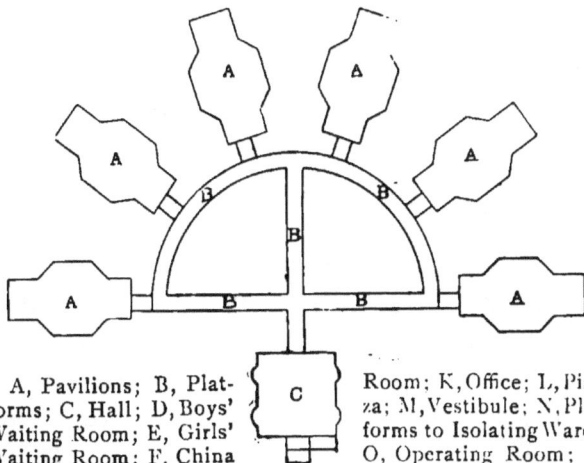

A, Pavilions; B, Platforms; C, Hall; D, Boys' Waiting Room; E, Girls' Waiting Room; F, China Closet; G, Officers' Dining Room; H, Dispensary; I, Sitting Room; J, Reception Room; K, Office; L, Piazza; M, Vestibule; N, Platforms to Isolating Wards; O, Operating Room; P, Etherizing Room; Q, Recovery Room; R, Toilet Pavilions; S, Lobby; T, Special Cases.

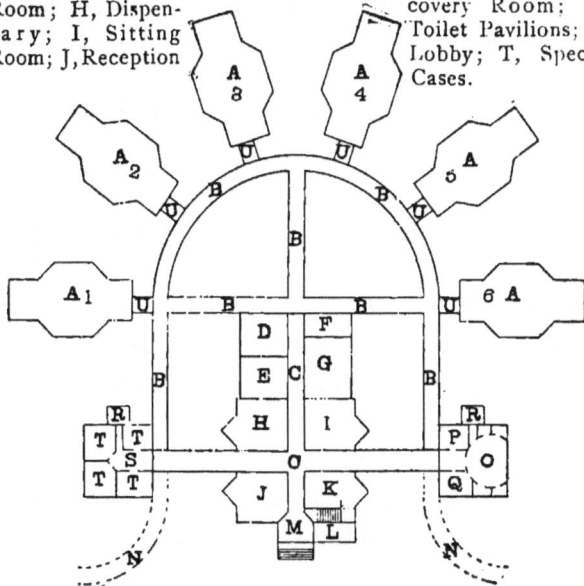

PLATE 10, PAGE 21.

Ohio Soldiers and Sailors Orphans' Home Hospital.
Xenia, O.

structure, costing less than $3.000, including all
plumbing and steam fitting. It is provided with a
solarium, and the ward contains beds or cribs for
twelve patients (children); it is heated by steam,
assisted—for ventilating purposes—by four fire
places, and four registers in the ceiling, above which
is a large air space under the roof. An illustration
of it also is shown on plate 11, page 23. Provision is
also made for a special room, and a nurse's room in
connection with it and the ward; the special room
is not intended for contagious diseases, but for an
interesting case or one which demands isolation.
Toilet arrangements are not arranged in a separate
pavilion, although there is a lobby between them
and the ward; a small closet is provided for physicians' use, and a closet, which has a window in it,
opens out of the ward. Although it is not especially
intended for linen, it can be used for that purpose.
The whole structure is connected to a permanent
building belonging to the U. S. Government operated by the Lakeside Hospital Corporation until the
end of a nearly expired lease, when it will probably
be removed. Therefor the cost was kept at a minimum and the excellent results of the heating and
natural ventilation are all the more gratifying for
that reason. An account of the results of a test by
the writer, showing the ventilation in this ward, is
given in the chapter on the renewal of air.

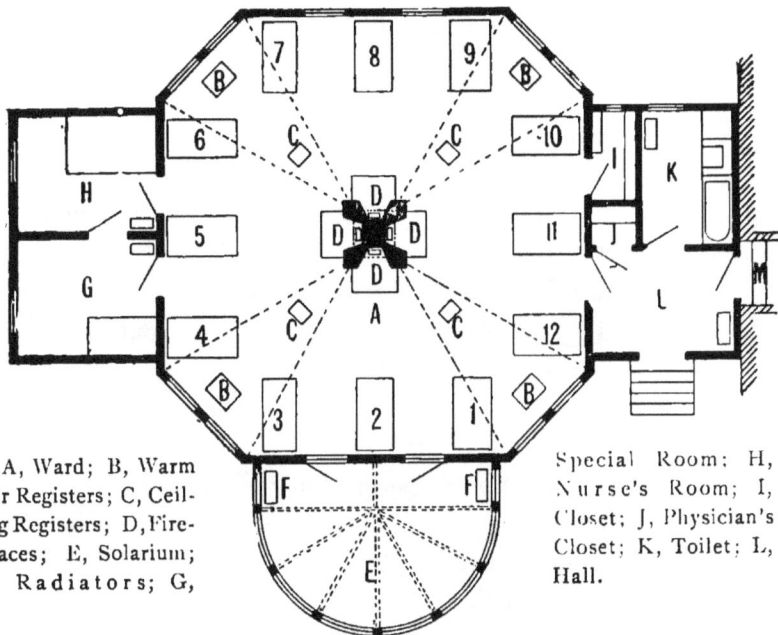

A, Ward; B, Warm Air Registers; C, Ceiling Registers; D, Fireplaces; E, Solarium; F, Radiators; G, Special Room; H, Nurse's Room; I, Closet; J, Physician's Closet; K, Toilet; L, Hall.

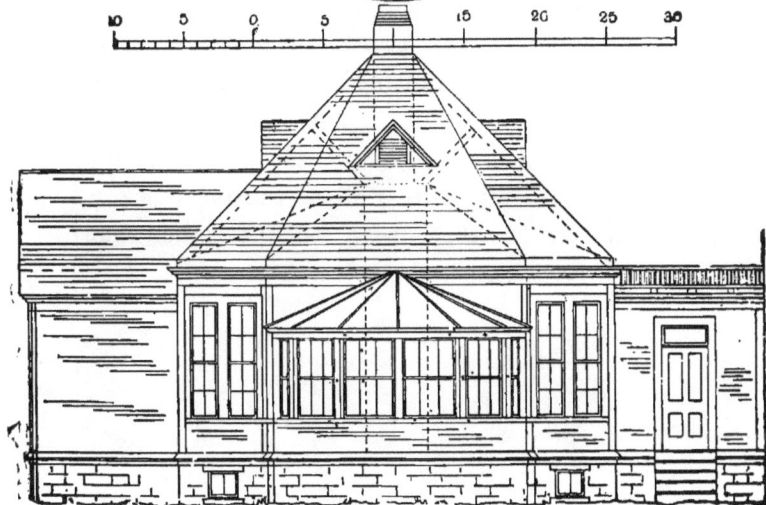

PLATE 11, PAGE 23.

Temporary Ward, Lakeside Hospital, Cleveland.

CHAPTER IV.

SKELETON PLANS.

Almost the first thing to be decided after determining the shape and character of the pavilious, is of course, the arrangement of the outline or skeleton of the hospital.

Many forms have been used, and it may be interesting to note some of the typical outlines which have been given them.

No attention has been paid to the precedence of the skeleton plans in this work, either as regards the date of erection or permanency of the buildings. A further description of some of these hospitals may be found in another chapter.

On some of the plates where more than a mere outline is shown, the black portions denote two or more stories of height: scored portions represent one story enclosed: outlined portions represent open or enclosed galleries, corridors and platforms. As some of the buildings do not now exist, having been temporary structures, all of the departments are not lettered, as the general outline rather than detailed information is the object.

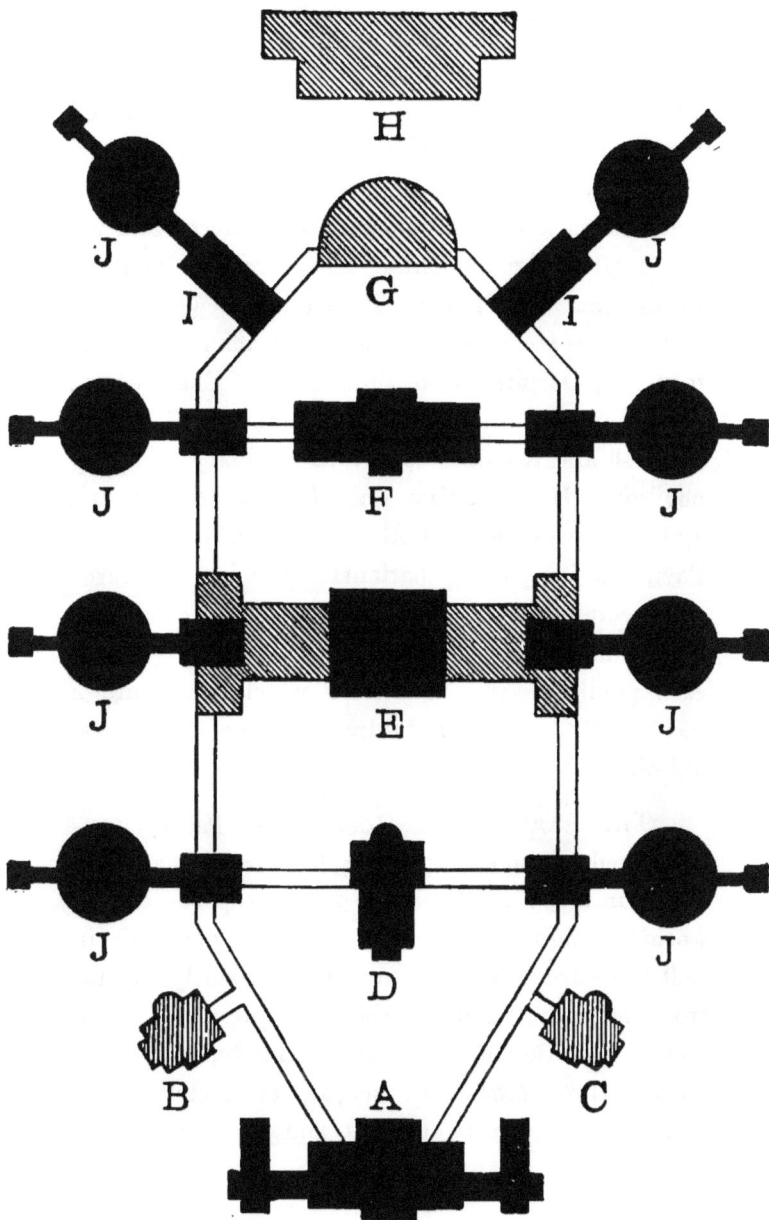

PLATE 12, PAGE 25.
Skeleton Plan, Antwerp Civil Hospital.
See also plates 7 and 8, pages 15 and 17.

THE ANTWERP CIVIL HOSPITAL, of which an enlarged plan has been given on preceeding pages, is shown on page 25, plate 12.   The letters on the plan indicate departments as follows. A, Administration building containing offices and administrative departments. B, Operating Pavilion. C, Morgue. D, Chapel. E, Kitchen, sewing room, store rooms etc. and Pharmacy and Chemical Laboratory. F, Dormitories of the Sisters of Charity who act as nurses in this institution. G, Russian, Turkish and Roman—as well as ordinary—bathing pavilion. H, Laundry and Engine rooms and appurtenances, with Boiler room underneath. I, Pavilions for paying patients. J, Circular wards with rectangular "head houses," and toilet pavilions. The total number of circular wards is 16, (two in each pavilion) in addition to which there are 4 rectangular wards for children in one of the central buildings.

THE LARIBOISIERE HOSPITAL of Paris, as represented on page 27, plate 13, has served as a prototype for many hospitals in different parts of the world, though it is claimed that it is defective in that the ward pavilions are not sufficiently distant from each other, and that the stairs prevent a satisfactory isolation of ward air. The buildings are mostly three stories in height, connected by a gallery and one story sections, and enclosing a large

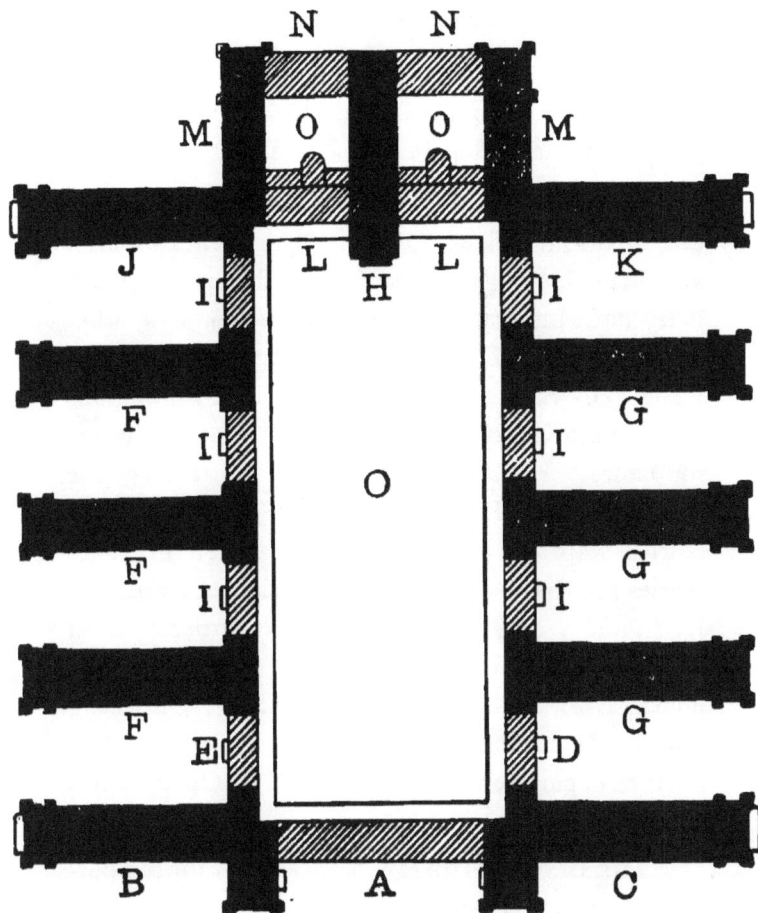

PLATE 13, PAGE 27.

Skeleton Plan, Lariboisiere Hospital, Paris.

open court 150 feet by nearly 400 feet. At the rear there are also two smaller courts. The letters on the plan indicate departments as follows.

A, Administrative departments. B, Kitchens etc. on ground floor, Officers' apartments on first and Attendants' rooms on second floors. C, Pharmacy and Laboratories on ground floor and other stories like B. D, One storied Reading Room for males. E, Same for females. F, F, F, Pavilions of three stories for female patients. G, G, G, Same for males. H, Chapel and vestry, containing the tomb of Mme. Lariboisiere. I, I, I, I, I, I, One storied wards or day-rooms. J, Pavilion of three stories for the Sisters of Charity, who act as nurses. K, Laundry and appurtenances on ground floor, Linen rooms on first and employees' rooms on second floor. L, L, Baths for each sex. M, M, Operating rooms. N, N, Morgue and dissecting rooms, Ambulance department etc. etc. O, O, O, Open courts and gardens.

THE LINCOLN HOSPITAL, was one of the temporary structures erected by the U. S. Goverment during the Civil war, and was located in Washington. A skeleton drawing of it is shonw on page 27, plate 13, fig. 1. It was shaped like the letter V and consisted of a series of wards connected by covered, but not enclosed, platforms forming an angle of about 50°. The chief criticism to be made of the

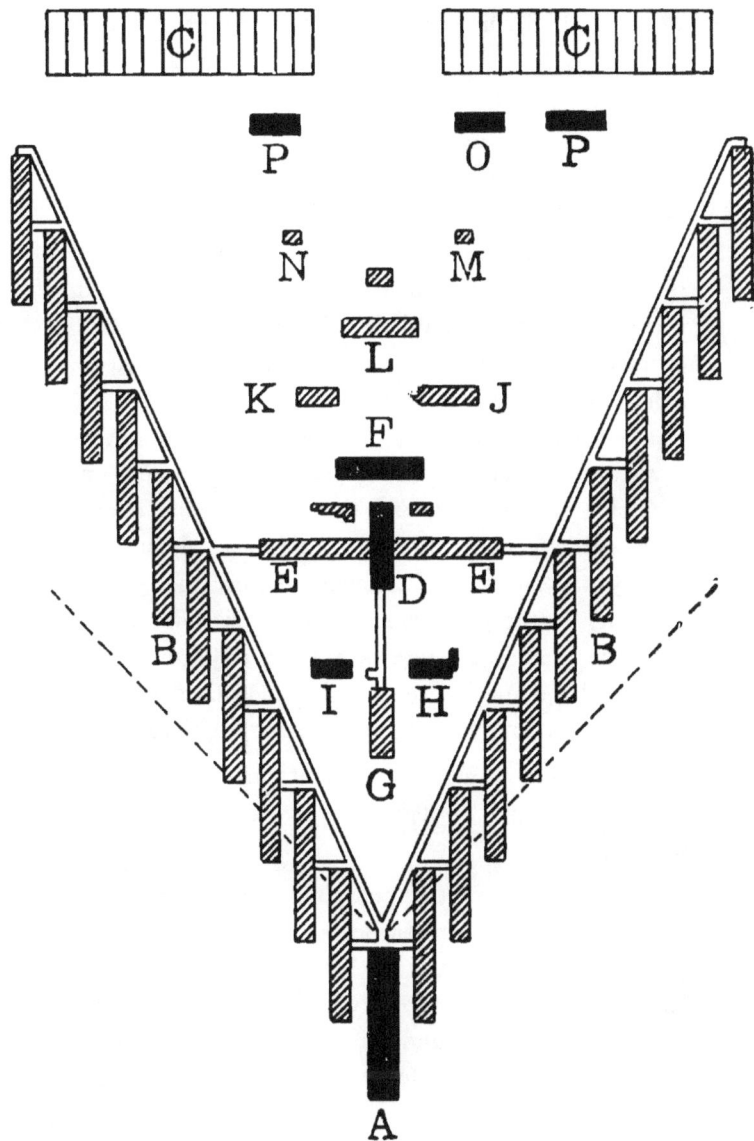

PLATE 14, PAGE 29.

Skeleton Plan, Lincoln Hospital, Washington.

Erected for temporary use during the Civil War.

arrangement is that the wards were not far apart, but the mortality does not appear to have been greater than in similar hospitals of different form. The same length of arms spread to an angle of 90° as shown by dotted lines would have added over 50 per. cent. to the distance between the wards, without increasing the cost of construction. Each ward was constructed of rough boards, whitewashed; roofs covered with tarred paper, and the walls plastered to a height of 8 feet from the floor. There were 20 wards, each 24 feet wide, 187 feet long, 16 feet to the eaves and 20 feet to the ridge, containing 31 beds on each side of the ward, and therefor allowing about 1,300 cubic feet of air space to each bed. Food was conveyed to the wards by means of a railroad 2,156 feet long on the platforms. The capacity was 2,575 patients, including those occupying adjoining barracks and 100 hospital tents which had raised wooden floors, where many of the worst cases were treated.

The total number of patients treated in this hospital during its operation was nearly 24,944, of whom 1,221 died.

The letters on the plan indicate departments as follows. A, Administration. B, B, The two lines of wards. C, C, Tents. D, Kitchen. E, E, Dining Rooms (total seating capacity 860). F, Commissary Department. G, Laundry. H, Sisters' Dormitories. I, Stewards' Quarters. J, Sutler. K, Chapel.

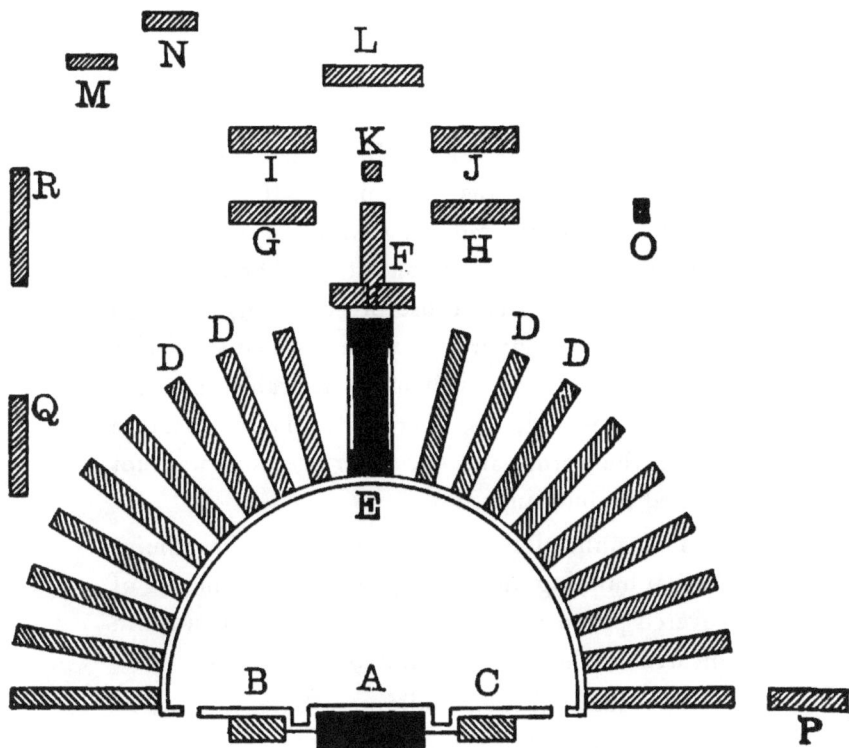

PLATE 15, PAGE 31.

Skeleton Plan, Hicks General Hospital, Baltimore.
Erected for temporary use during the Civil War.

L, Stable.   M, Dead-house.   N, Guard-house.   O,
Barracks.   P, P, Officers Quarters.

THE HICKS GENERAL HOSPITAL at Baltimore
page 31, plate 15, was built in a half circle, with
eighteen, one story, radiating wards; the buildings
were constructed of frame but of a rather more sub-
stantial character than those preceding it, although
it also was of a temporary nature: in many respects
its arrangement was much more desirable than the
Lincoln Hospital, as there was more distance
between the wards, and the interior court was not
occupied by buildings.

The connection between wards and the administration building and dining room was by means of
a covered platform.   The different departments are
indicated on the skeleton plan as follows ;—

A, Administration building.   The first story
contained offices for the surgeon and other execu-
tive officers, and a library and printing office: on
the second floor were sleeping apartments for them.
B, Linen room.   C, Dispensary and operating room.
D, D, Wards radiating from the covered platform.
E, Dining room: the chapel and nurses' dormitories,
which were above it, were reached by outside stairs.
The dining room was capable of seating 1,200
patients.   F, Kitchen, laundry, engine and boiler
room.   G, Special dormitory for detailed men.   H,
Knapsack room, where the property of patients was

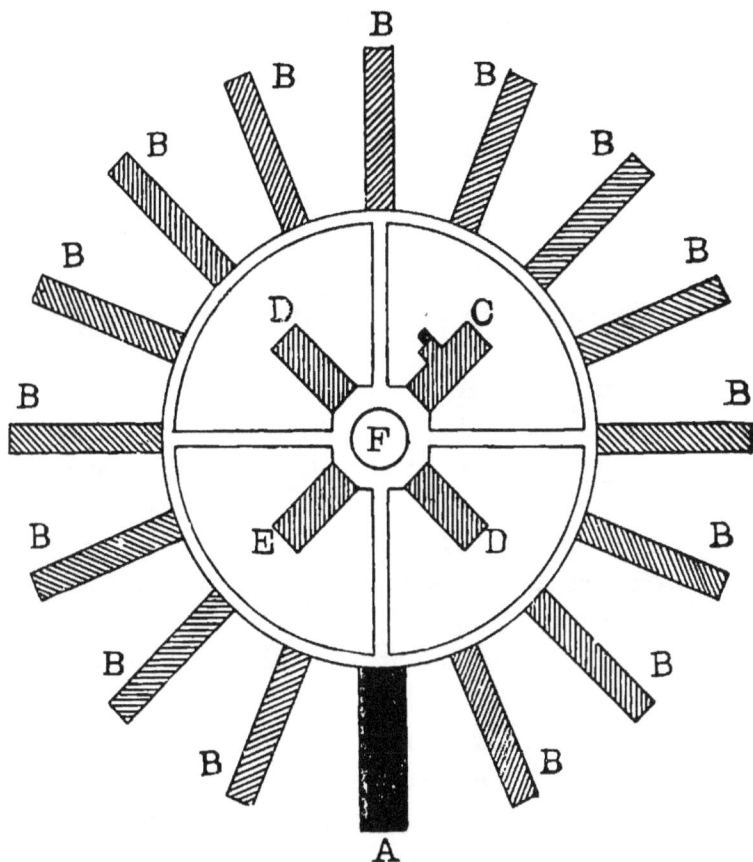

PLATE 16, PAGE 33.

Skeleton Plan, Sedgwick, Hammond and other similar Hospitals.

Erected for temporary use during the Civil War.

kept. I, Commissary store-department. J, Quartermaster's store-department. K, Tank supplying water to the buildings. L, Dormitory for guard. M, Stable. N, Wagon house. O, Sutler. P, Steward's quarters. Q, Workshop. R, Isolation ward for contagious cases.

In addition to this department there were at different points on the ground (outside of the circular portion,) accommodations for the officers; also the guard room and entrance lodge. The buildings were lighted by gas and supplied from city water works; the tank was used for the purpose of a water supply in case of fire. Each ward, at its further extremity, was supplied with a lavatory, bath room and water closet. The building was not opened for patients until about the middle of 1865, or the close of the war.

THE HAMMOND HOSPITAL was one of a series of similar buildings erected for temporary hospital uses during the war, and was in many respects superior to either of the forms we have just examined, in that the distance between the pavilions was greater than in the foregoing examples. Page 33, plate 16.

It consisted of a covered platform forming a complete circle with four platforms radiating from the centre of the circle, with wards radiating from the circular platform. There were fifteen one story pavilion wards 24 feet wide by 145 feet long, and an

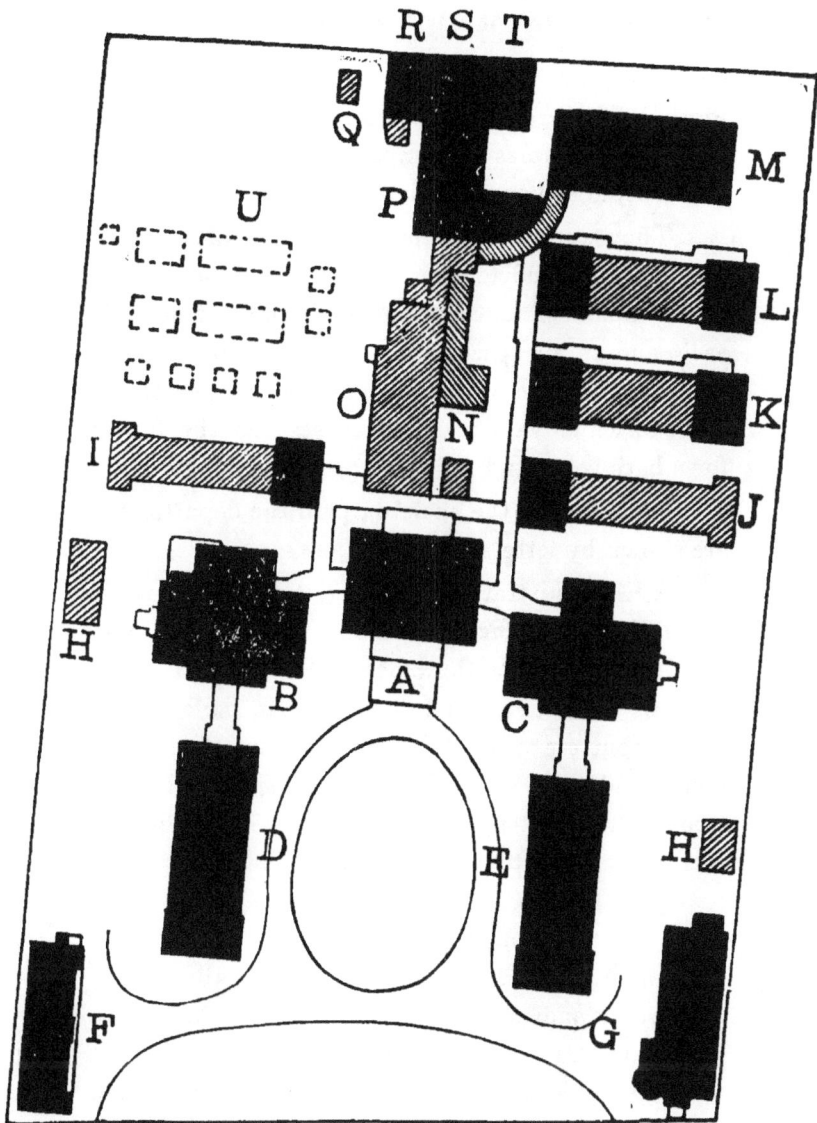

PLATE 17, PAGE 35.
Skeleton Plan, City Hospital, Boston.

administration building 40 feet wide and 145 feet long. The wards were very similar to those of the two first examples of temporary structures, being constructed of boards with shingled roofs.

They were, like the preceding examples, raised above the ground to allow a free circulation of air under them. The wards had a capacity of 40 beds each, and an air space of about 1,200 cubic feet for each bed. There were two small rooms at each end, three of which were used for nurses and the fourth for a bath room and toilet purposes.

Referring to the skeleton plan the departments are shown by letters as follows;—

A, Administration building, two stories high, containing the office and dispensary on the first floor, and rooms for the officers above it. B, B, B, Wards. C, Kitchen and engine room, boiler room, store room etc. D, D, Dining rooms. E, Guard house, and store rooms for knapsacks, etc. F, Cistern containing 150,000 gallons. The accommodation for patients in the fifteen wards was 600.

THE BOSTON CITY HOSPITAL consisted when erected, of five semi-detached buildings exclusive of the working service department. An outline of it, in 1891, is shown on page 35, plate 17. Since its enlargement in 1876, several one story pavilions have been erected, one of which (I) is shown on page 37, plate 18, being the one in which occurred the

A, Ward; B, Day Room; C, Bath; D, Water Closets; E, Dining Room; F, Nurse; G, Special Ward; H, Stairs; I, Patients' Wardrobe; J, Linen Room; K, Connecting Platform; L, Steam Coils for warming fresh air; M, Steam Coils to assist ventilation.

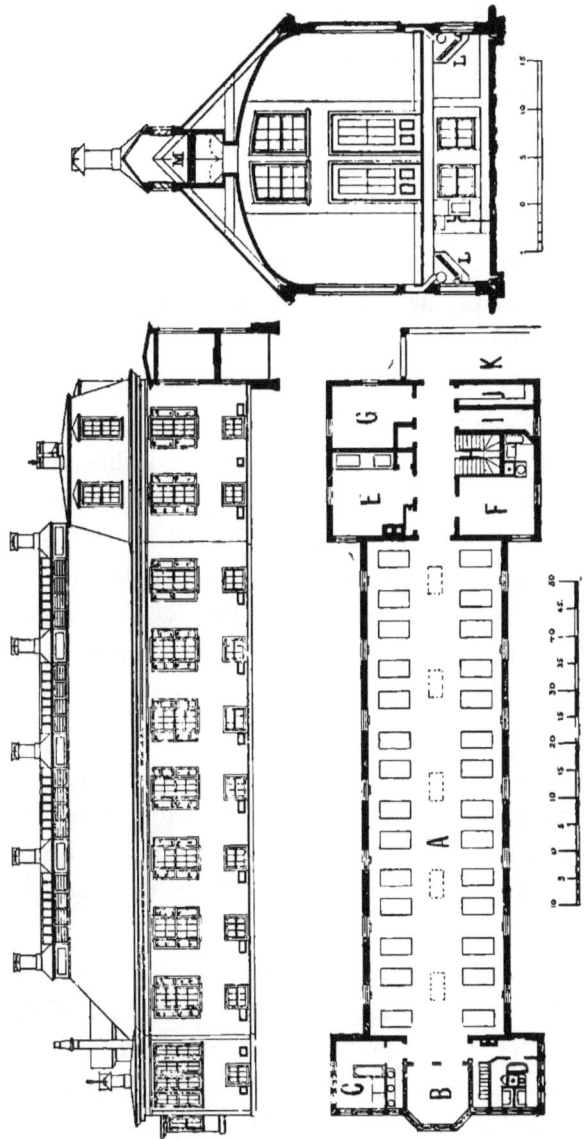

PLATE 18, PAGE 37.

Surgical Ward, City Hospital, Boston.

interesting experiments of Dr. Edward Cowles on the subject of ventilation.

This hospital, as first erected, was one of the examples (more often found in old than in recent buildings,) of architectural effect out-ranking the accommodations for patients. Therefore wards were arranged in the attics and basement, although the latter have since been given up, and the former are not used to any extent in warm weather.

While not as pretentious as many of the more recent structures, it is certainly admirably arranged in one respect, viz., the complete separation of surgical and medical cases. It will be noticed that all surgical cases—either male or female—are on the left of the administration building, while the medical cases are on the right: this arrangement extends even to the Out Patient departments.

The surgical and medical attendants are totally separated from each other, when on duty, so that there is no danger of any infection being spread from a medical to a surgical case, or *vice versa*, as might happen if one staff of officers were obliged to attend all cases indiscriminately.

The chief criticism to be made of this hospital is that the Medical, Scarlet Fever and Diphtheria wards, (J, K, L,) are rather too near each other. It might be inadvisable to erect more surgical pavilions without removing the kitchen, bakery, refriger-

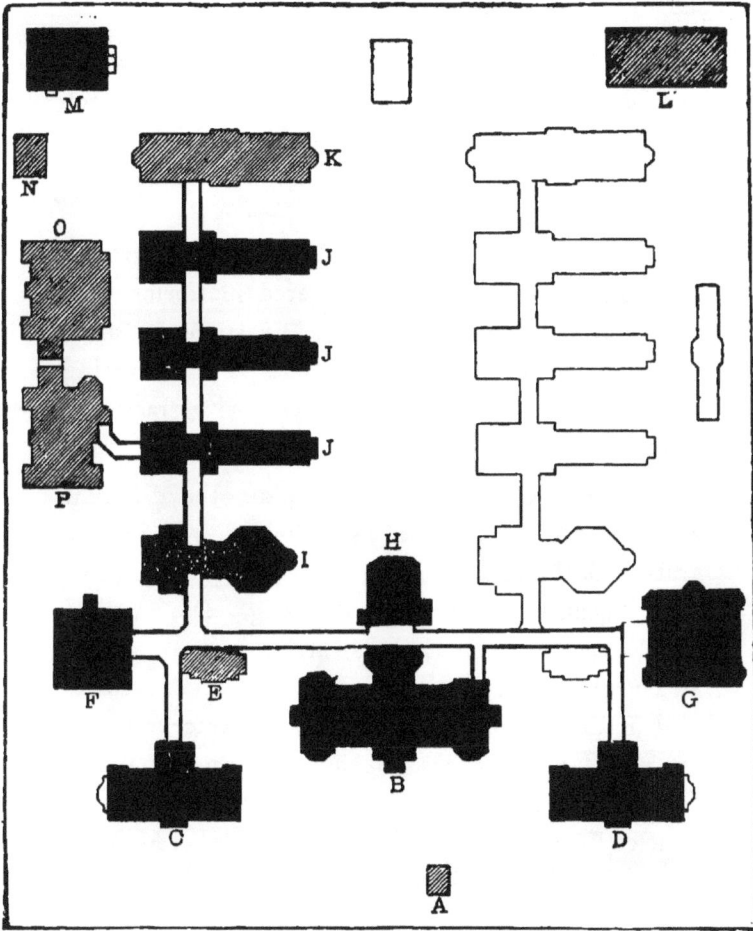

PLATE 19. PAGE 39.

Skeleton Plan. The Johns Hopkins Hospital, Baltimore.

ator and greenhouse to the rear of the lot, which would admit of a open court 125 feet wide between the ends of the one story medical and surgical pavilions. A new greenhouse might be erected in the centre. Communication between the kitchen and each pavilion could be made by an under-ground passage, as well as by the covered platforms, and food would be quickly conveyed in tram-cars as at present. An extensive use of tents is made during the summer season, for serious cases, with gratifying results.

Referring to the skeleton plan, the different departments are indicated as follows. A, Administration Building. B, Surgical Building and Operating Theatre. C, Medical Building. D, Surgical Pavilion, 3 wards. E, Medical Pavilion, 3 wards. F, Surgical Out Patient department. G, Medical Out Patient department (see also chapter on this subject). H, H, Horse Sheds. I, Surgical Ward, (see also page 37, plate 18). J, Medical Ward. K, Scarlet Fever Ward. L, Diphtheria Ward. M, Male Isolating Ward. Female Isolating Ward above it. N, Greenhouse. O, Kitchen and Bakery. P, Laundry. Q, Garbage. R, Morgue. S, Boiler House. T, Carpenter Shop. U, Tents.

THE JOHNS HOPKINS HOSPITAL of Baltimore is considered the *ne plus ultra* of the modern hospitals of the United States to-day, but its great cost as

PLATE 20, PAGE 41.

Perspective of The Johns Hopkins Hospital, Baltimore.

compared with its capacity will prevent for years to come, if ever, a repetition of all its successful features. The plan was decided upon by the trustees after consultation with a number of well known physicians who had made a study of hospital arrangements, and the work was completed under the most vigilant supervision.

The buildings are of brick with stone trimmings, and while every care was taken to secure the best sanitary results, the external appearance was not neglected. The skeleton plan on page 39, plate 19, shows the buildings as erected, and in faint lines, the plan as projected.

The various departments are indicated as follows. A, Gate Lodge. B, Administration. C, Male paying patients. D, Female paying patients. E, Bathing Pavilion. F, Kitchen, Bakery and Boilers. G, Nurses' Home. H, Apothecary, (hospital dispensary) with officers' dining room and servants' rooms above. I, Octagon Ward. J, J, J, Common Wards. K, Isolation Ward. L, Laundry. M, Pathological Building. N, Stable. O, Dispensary, (out patient department). P, Amphitheatre.

The basement or ground floor of nearly all of these buildings is on a level with the ground, and is practically unused. The floor of the connecting and enclosed corridor, indicated by heavy lines, is also on the same level, with a flat roof which is on a

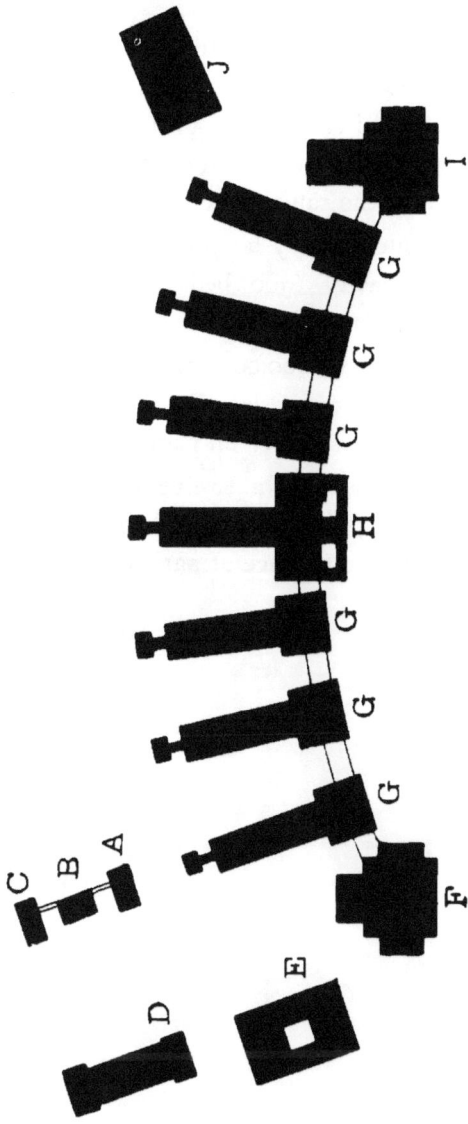

PLATE 21. PAGE 43.

Skeleton Plan. San Andica Hospital, Genoa.

A, Fuel; B, Gardener's House; C, Upholstering Department; D, Laundry; E, Mortuary, Dissecting Room and Museum; F, Student's and Lecture Room; G, Wards; H, Chapel; I, Administration Building; J, Convalescent Paying Patients.

level with the first floor of the wards, and is used as a terrace connection between them. It will be noticed that this hospital plan, as projected, differs from those preceeding it, in that all the wards would, if thus built, face in the same direction.

Probably the most costly hospital in the world, considering its capacity (300 beds), is the San Andrea hospital of Genoa, Italy.

Its cost was over $2,400,000, and it was built through the generosity of the late Duke of Galliera. The skeleton plan of it is shown on page 43, plate 21.

The six pavilions contain twelve wards of 20 beds each, while a large infirmary ward and separate rooms accommodate the balance of patients.

CHAPTER V.

CLASSES OF HOSPITALS.

SPECIAL HOSPITALS. Many small hospitals in large cities cover a wider field of treatment than is desirable, but the tendency seems to be towards the establishment of special hospitals for the segregation of certain classes of cases.

This is one of the natural results of specialism in medical practice, and its advantages are manifold. The specialist is enabled to secure to his patients the attention of nurses educated to the requirements of his special work. There is little danger of infection from cases of a different character. The hospital may be located in a city, on a comparatively small lot, easily obtainable, instead of necessitating acres at a greater distance from the center of population.

As all hospitals are usually private enterprises, the dislike, so common among the better class of people, of "going to a hospital" is lessened : there they are protected from the well-meant but often irritating visits of friends, and the physician can give them better attention than if they are scattered about the city in their homes.

Such hospitals would probably be limited to three or four classes, viz: one for Cancer patients; one for Gynaecological cases; one for "Lying In" cases; and perhaps one for Opthalmic and Aural cases.

As population increases these would increase in size, until what might have had its commencement in a city dwelling house, would be occupying buildings especially erected for its use. The first two classes of patients mentioned above, while not dangerous to others in a general hospital, are themselves particularly sensitive to infection; for this reason special hospitals (or at least the most careful isolation), is advised for them.

If knowledge of disease has been increased by the labors of specialists, why should not the treatment of disease be improved by special hospitals?

Notwithstanding the advent of anti-septic surgery, the decrease of mortality percentage, and the greater infrequency of septicaemia than heretofore, what an advantage it will be when the erection of hospitals for special work of the above character will be the *rule* rather than the *exception*.

GENERAL HOSPITALS. While it is desirable that the special cases referred to above should be treated in separate buildings, there is no question

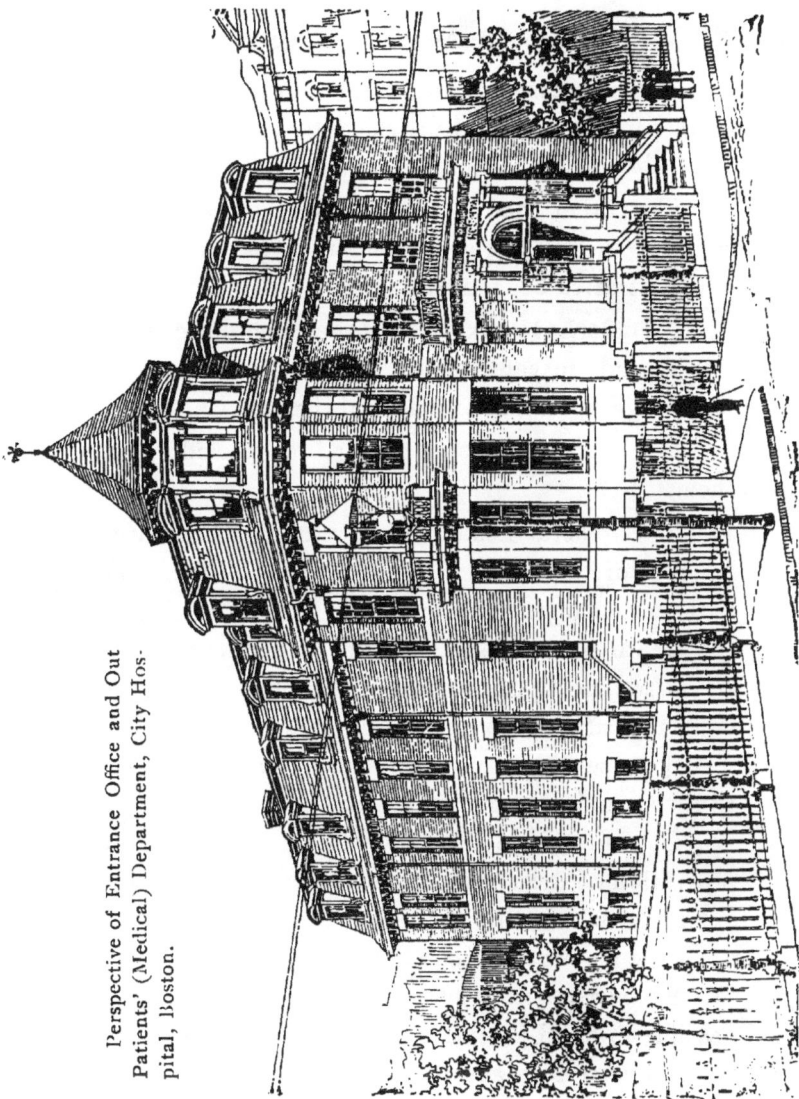

Perspective of Entrance Office and Out Patients' (Medical) Department, City Hospital, Boston.

but that, if necessary, they can be successfully treated in general hospitals provided they are detached, or semi-detached, from the main group of buildings.

The other medical and surgical cases can be safely treated in one group of buildings, under one administrative department.

A detached auxiliary group should contain wards and private rooms for the treatment of Measles, Scarlet Fever and Diphtheria.

One-story pavilions should be provided for them.

The general hospital will remain an essential feature of medical practice for all time, and anti-septic surgery and the constant modification of the treatment of disease, under the researches of *specialists* mostly, render the treatment of medical and surgical cases in one building much safer than was the case not many years since, though separate wings should be provided for each.

Of course care should be taken to render the separation of wards as complete as possible, and the extent to which this can be done depends largely upon the amount of land obtainable.

HOSPITALS ON CITY LOTS.   But it is not always possible to build extended pavilion hospitals of only one or two stories, and when this cannot be done it is necessary to arrange a more condensed plan ; for rather than to have no hospital at all, one planned

A, Office; B, Reception Room; C, Private Waiting Room; D, Admitting Room; E, F, G, Eye and Medical Diseases; E, Examining; F, Consulting; G, Testing and Examinations; H, General Waiting Room; I, Stairs to third floor only.

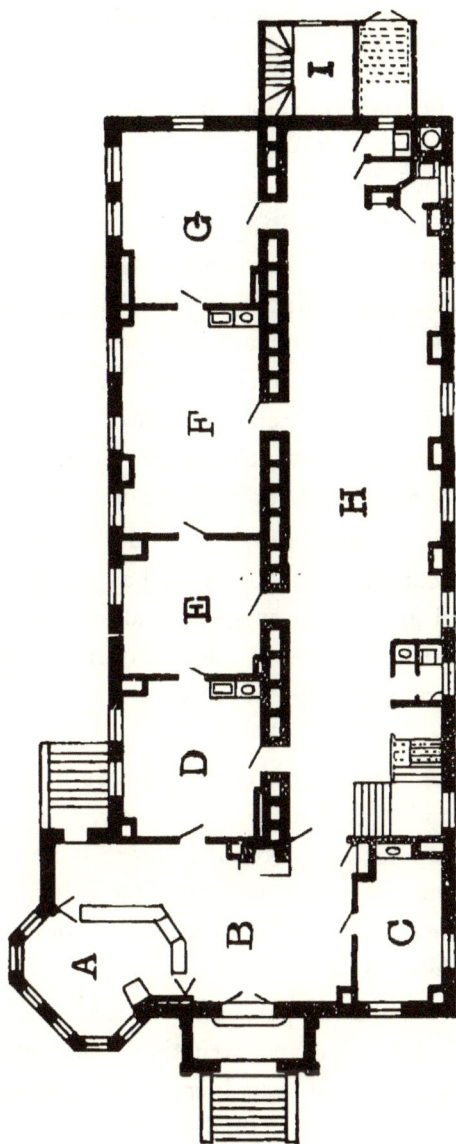

PLATE 23, PAGE 49.

First Floor Out Patients' (Medical) Department. City Hospital. Boston.

with an earnest endeavor to obtain isolation of wards under one roof must be the alternative. Witness the successful working of the New York Hospital on Fifteenth Street. Pages 55, 57 ; Plates 26, 27.

It was after visiting this building, and while visiting another of four stories, that the celebrated Prof. Lister, of Edinburgh, of anti-septic dressing fame, said : "It is immaterial how many stories of wards there may be in a hospital, provided that the details of the anti-septic method are accurately carried out in all of them. If these details are faithfully observed hospitalism can be prevented."

What was then (1876) considered experimental by some conservative operators, has now become the universal practice in surgery, and with the revolution of surgery came a modification of the ideas then becoming popular concerning low hospital buildings.

It has been found that in wards of one story, with abundance of sunlight and air, the results of surgical operations were much more successful than in the many-storied, old and saturated hospitals which the one-storied pavilions were commencing to supersede.

Thus while the pavilion remained the accepted standard of ward construction, the anti-septic method admitted of even increased height of buildings, should a limited site demand it.

J, K, Examining Rooms; I, M, N, Diseases of Nervous System; L, Treatment; M, Testing; N, Consulting; O, P, Diseases of the Throat and Eye; A, Stairs from first to third floors; R, Waiting Room; S, Diseases of Women.

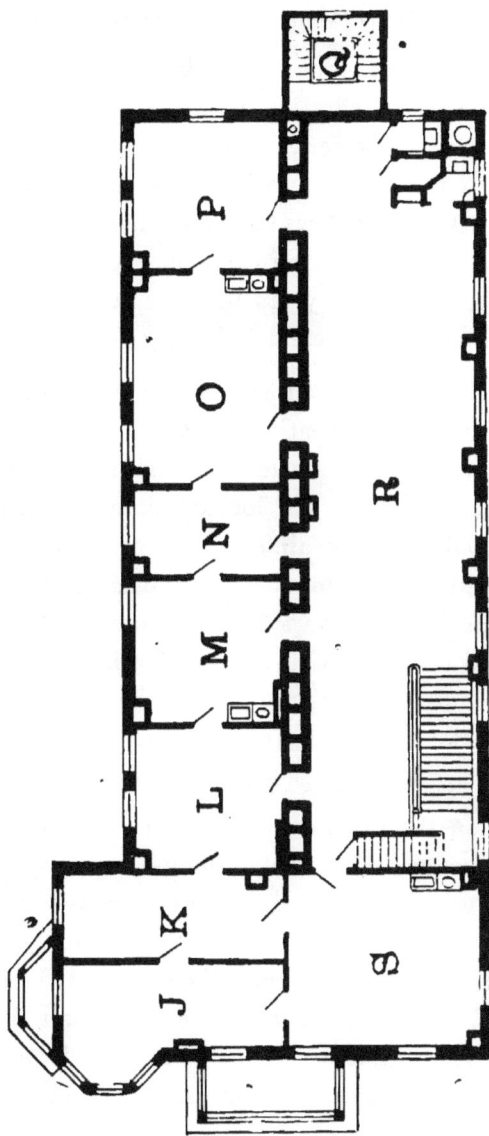

PLATE 24, PAGE 51.

Second Floor Plan, Out Patients' (Medical) Department, City Hospital, Boston.

It also remained an acknowledged fact, that while such comparative immunity from Hospitalism was *possible*, the fewer-storied buildings were beneficial to all patients: owing to the decreased expense of constructing low buildings and the increased amount of sunlight gained in a given space thereby, they still, and for years to come, will be built and *should* be built when sufficient ground is obtainable.

One of the advantages of the fan for mechanical ventilation and heating, lies in the fact that in hospitals of many stories the fresh warm air may be inspirated by one fan located in a sanitarily safe place, and vitiated air expired in another, and if the ventilating fan does its work constantly the tendency to cause a vacuum will prevent the air from one ward mixing with that of another by means of the windows, which might be the case if natural ventilation only were relied upon.

"OUT PATIENTS" DEPARTMENT. In large hospitals on the pavilion plan, where the grounds are extensive, this department should be an *out* department, entirely isolated excepting, possibly, an underground passage from the other buildings. In a hospital on a city lot, this, of course, may be impracticable, but it should in that event be all the more carefully ventilated. It is one of the advantages of the fan ventilating apparatus, that divisions of a hospital requiring more rapid ventilation, or more

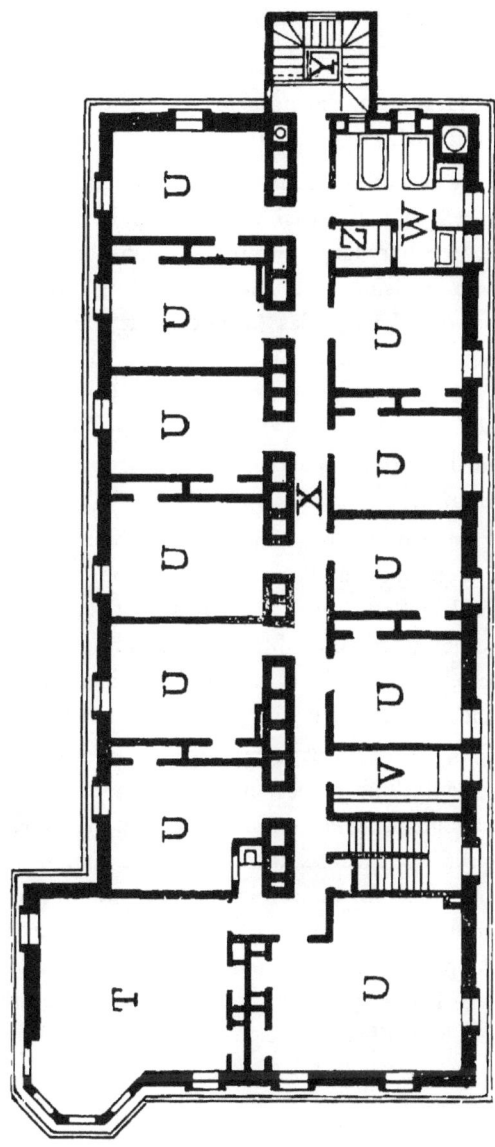

T, Sitting Room, Male Nurses; U, Bedrooms, Male Nurses; V, Linen Closet; W, Toilet and Bath Rooms; X, Hall; Y, Stairs.

PLATE 25, PAGE 53.

Third Floor Plan Out Patients' (Medical) Department, City Hospital, Boston.

frequent renewal of air than others, may be arranged with a larger ventilating shaft. This can be readily accomplished.

The reception room of these patients, if space is limited, may be used in common by males and females; otherwise it would be advisable to have separate divisions for them. Leading from this room should be a series (according to the probable number to be treated) of examination rooms, which need not be more than eight by ten feet square if the space is limited. Toilet facilities should be provided for this department, to insure its complete isolation.

Pages 59, 61, 63; Plates 28, 29, 30, show the department for Out Patients at the Johns Hopkins Hospital in Baltimore. The Boston City Hospital has recently completed a new building for "out patients," which will be used for the examination of medical cases exclusively.

It is shown on pages 47, 49, 51, 53; plates 22, 23, 24, 25. The objection to French or Mansard roofs for hospitals does not apply so particularly to this building, because the story in the roof is occupied by rooms for male nurses, and patients would probably never be placed in any of these rooms. This department is so complete and the arrangements of it are so perfect, that it is given as one of the best which has yet been erected. It is of brick

with stone trimmings, and is 100 feet in length and an average width of about 40 feet. The ventilation and heating is by the natural method, and the warm air is supplied to all the rooms by indirect radiation at the outside walls; the central longitudinal wall is divided into flues and serves for ventilation. The air in these flues is warmed (and thereby given an upward current) by steam pipes which are connected with the radiators in the Mansard story. The former "out patient" department is to be used in the future as a *surgical* "out patient" department. This new building will also serve as an entrance office for the hospital.

EMERGENCY HOSPITALS. One of the chief hospital needs of almost all cities is for emergency hospitals to which purely surgical cases demanding immediate attention can be quickly transported. How often has the distance to the hospital caused the death of some unfortunate through the slipping of the hastily applied compress, or the continuance and intensification of the collapse attending the accident.

In an emergency hospital devoted entirely to "accident cases," there would not be the unspoken fear on the part of surgeons that their efforts were being counteracted by dangerous medical cases in adjoining or neighboring wards.

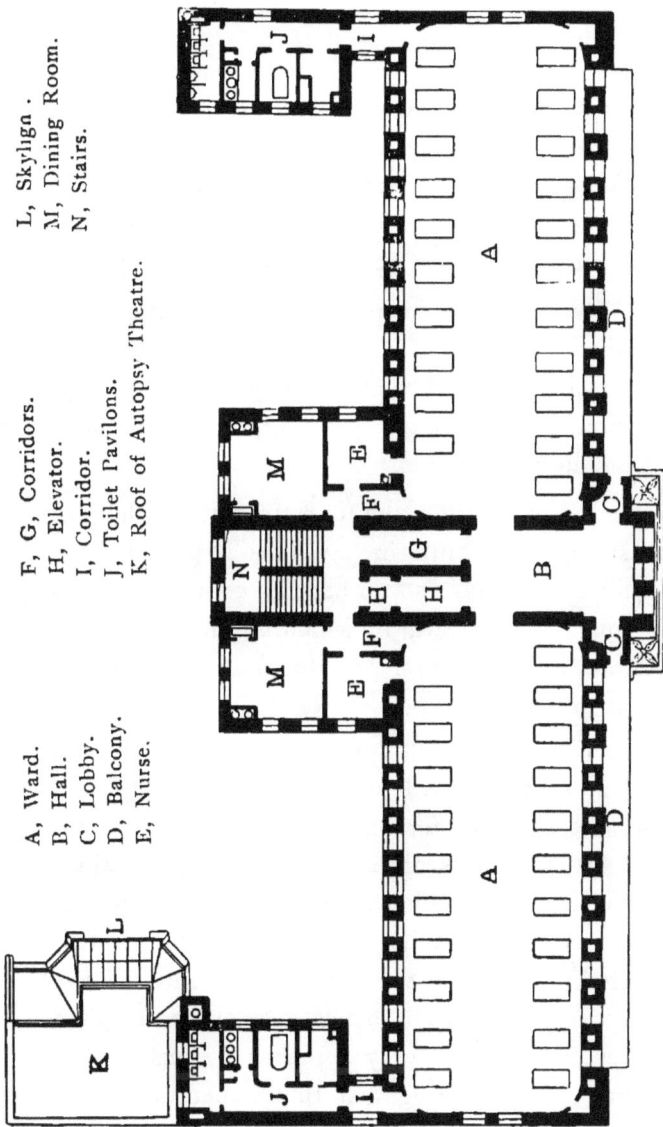

A, Ward.
B, Hall.
C, Lobby.
D, Balcony.
E, Nurse.

F, G, Corridors.
H, Elevator.
I, Corridor.
J, Toilet Pavilons.
K, Roof of Autopsy Theatre.

L, Skylign .
M, Dining Room.
N, Stairs.

Plate 27, Page 57.

Second, Third and Fourth floors New York Hospital. Fifteenth Street.

An emergency hospital can be readily adapted to almost any lot and need not contain a large number of beds, though certain divisions are necessary for isolation of special cases, and for the sake of brevity reference is made to the proposed arrangements on pages 65, 67, 69, 71; plates 31, 32, 33, 34.

Of course *all* hospitals should contain an accident ward, so that when an emergency hospital is not near at hand the patient may receive prompt attention. Such a ward, however, should be built in a pavilion if possible.

In fact, the writer cannot help stating that *all* wards should be built in pavilions if possible. It would be desirable, also, to have in connection with it a convalescent ward, in which might be put mild cases, as well as several private rooms for patients who are able to pay.

TEMPORARY STRUCTURES OR HUTS. Walls of metal may be assumed to be useful only for small structures or huts for the isolation of the worst forms of contagious diseases, and even then their value is largely problematical. The cheerless appearance of such walls, usually of iron, unless plastered inside, with the added discomforts of the noise of the wind as it blows over the corrugated surfaces, must have a very depressing effect on the patient, and operate against that hopefulness and determination to get well that forms so powerful an

PLATE 28, PAGE 59.

Perspective of Out Patients Department, Johns Hopkins Hospital, Baltimore.

adjunct, with fresh air and sunshine, to the labors of the nurses and physicians.

Such structures are usually nearly square, consisting of a single room with a nurse's room connecting, and should seldom be arranged for more than two beds, preferably one, to secure the best results.

They are usually built on piers or cedar posts, with lattice work between them, and hinged sash, only one of which is shown in the perspective.

The sash should be kept open except in the coldest weather. A light skeleton of wood is used for walls and roof, over which the iron is securely fastened. The heating should be by means of a fire place in each room with a ventilating flue between the backs of the grates. Such a structure, but covered with clapboards or shingles, is shown on page 73, plate 35. The designation of the letters is, A, One of the sashes over the lattice ; B, Transom of window to open in ; C, Ward ; D, Nurse's Room ; E, E, Fireplaces ; F, Ventilating flue ; G, Porch.

Connected with the ventilating flue between the grates are pipes below the floor connecting with raised registers under the beds (H,) and also wall ventilators (I, J,) which connect into the ventilating flue between the ceiling and the roof.

The 16 x 16 ward contains, if constructed with a "half pitch" roof on 10 feet studs, 3,840 cubic feet or 1,920 cubic feet per. bed.

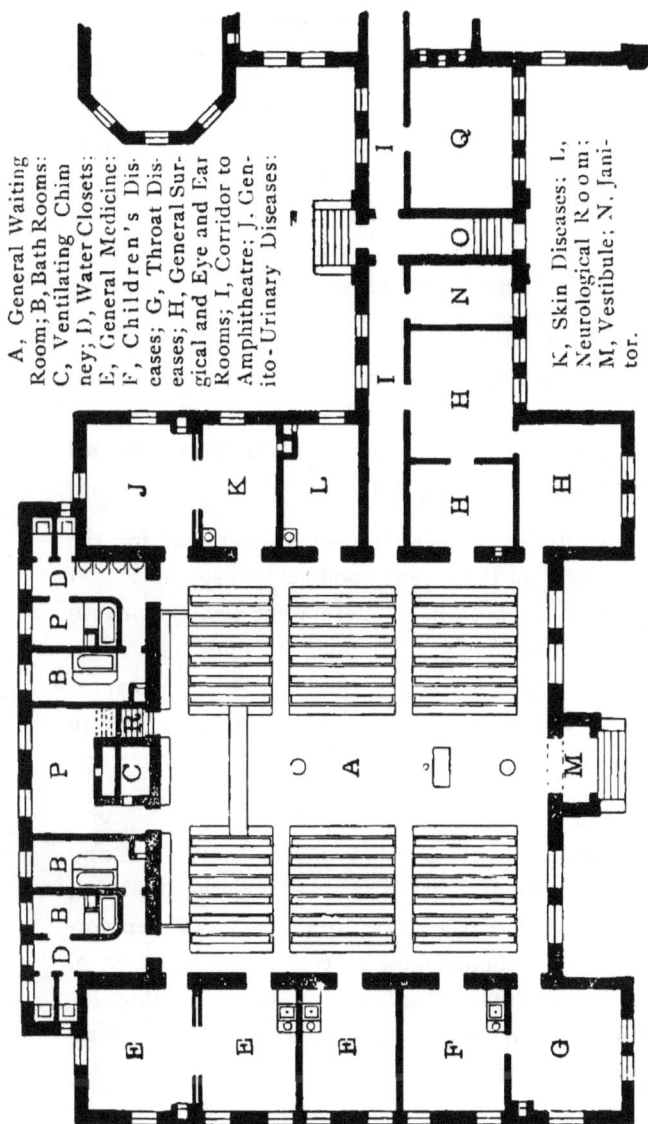

A, General Waiting Room; B, Bath Rooms: C, Ventilating Chimney; D, Water Closets: E, General Medicine: F, Children's Diseases; G, Throat Diseases; H, General Surgical and Eye and Ear Rooms; I, Corridor to Amphitheatre: J, Genito-Urinary Diseases:

K, Skin Diseases: L, Neurological Room; M, Vestibule; N, Janitor.

PLATE 29. PAGE 61.

Plan of Out Patients Department Johns Hopkins Hospital, Baltimore.

Windows should be placed on three sides of the nurse's room and on four sides of the ward. The doors should be glazed. A porch should be provided |for the benefit of convalescents. The chief objection to such huts is that the iron is practically indestructible, and might, even after the most thorough fumigation, prove fatal to future occupants.

It would seem preferable to erect such structures entirely of wood, and their isolated position would admit of their being torn down and burned after use for any length of time.

A number of cheap houses were erected for the sufferers from the recent flood at Johnstown, Pa., at a cost of $260 each. These were two stories in height. Previously a number of one story huts were put up at a cost, it was reported, of $125, for 10 x 20 feet in size, and $190, for 16 x 24 feet in size, o. b. c. Chicago, where they were made.

They were of course "portable" houses, viz: constructed in sections that would admit of taking down and removal, being fastened together by bolts, and are suggested here as a quick means of temporary relief from epidemics or contagious diseases in crowded asylums, "homes" and similiar State Institutions that frequently have no separate hospital pavilions provided.

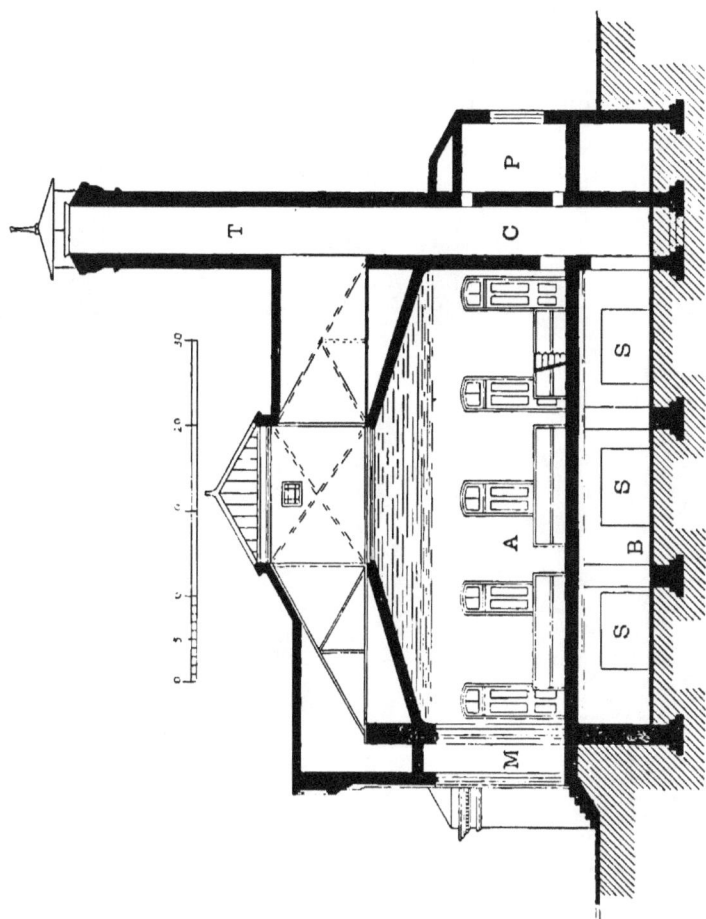

PLATE 30, PAGE 63.

Section of the Out Patients Department, Johns Hopkins Hospital, Baltimore.

A, General Waiting Room.
M, Vestibule.

B, Basement.
P, Pharmacy.

C, Ventilating Stack.
S, Heat Chambers.

T, Accelerating Steam Coils:

They could be erected rapidly after delivery of the sections on the ground, and the sizes above mentioned are usually carried in stock.

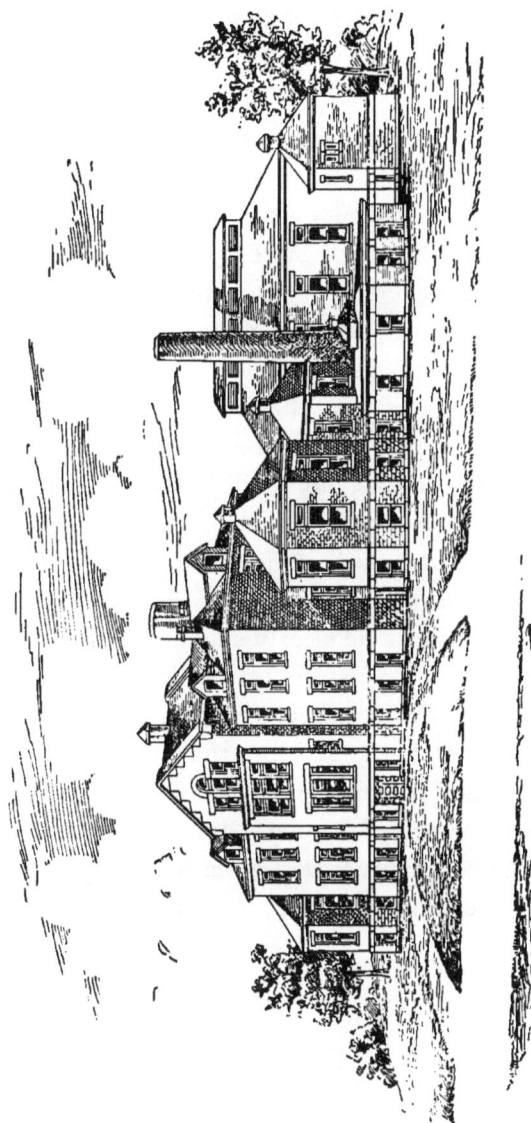

PLATE 31, PAGE 65.

Perspective of an Emergency Hospital.

CHAPTER VI.

U. S. MILITARY HOSPITALS.

In 1888 considerable improvement was made in the arrangement of Military or Post Hospitals, and outlines were published by the War Department as guides in planning such buildings whether erected for permanent or temporary structures.

The following description is taken from the general directions accompanying the outline plans:

"It is not expected that they will be literally followed. Circumstances of location, ground levels, nature of soil, climate, etc., will require variations from them to produce the best results at a reasonable cost; but when such variations are made, either in plan or material, the reasons for so doing and the estimated cost must be clearly stated in forwarding the detailed plans and specifications for the approval of this office. For example, these plans provide for cellars underneath the building to receive the heating apparatus. In some cases, owing to level of soil water, or to nature of foundations, dry suitable cellars cannot be constructed at a reasonable cost. In such case the walls of the basement may be made

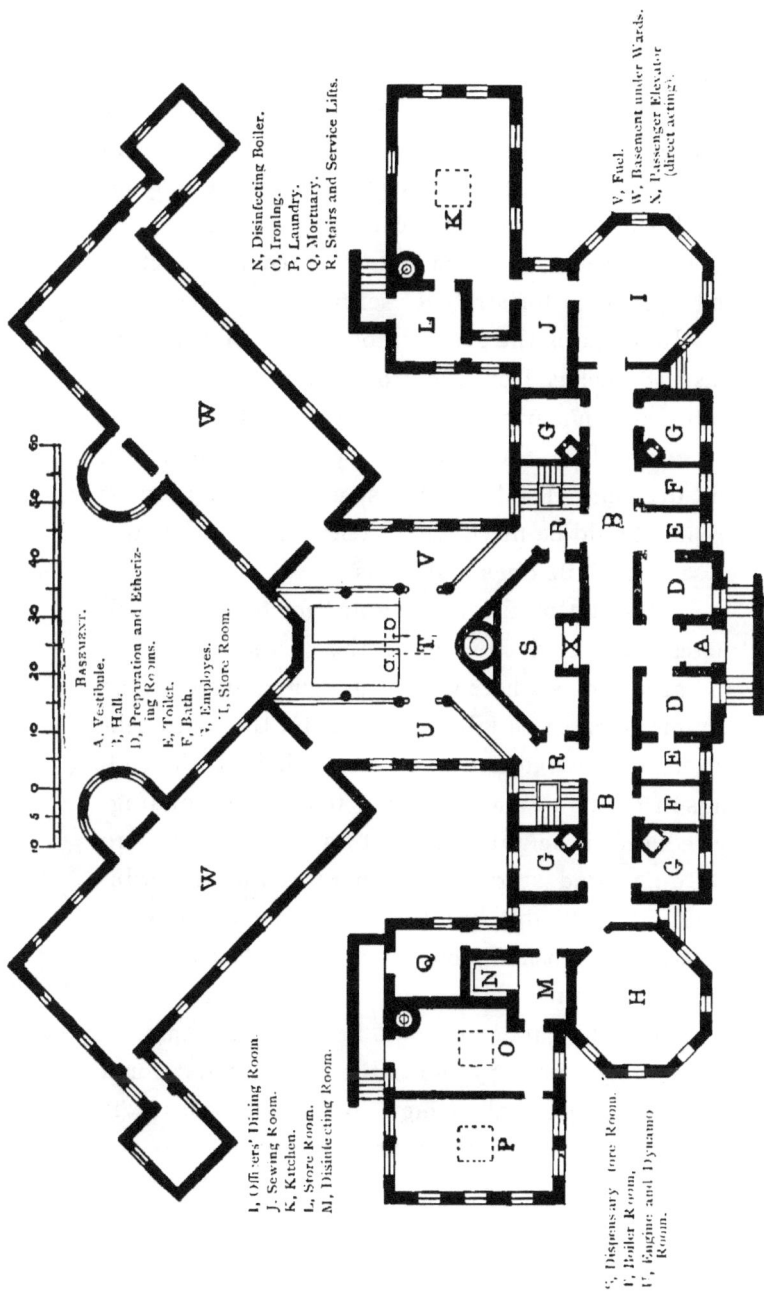

BASEMENT.

A. Vestibule.
B. Hall.
D, Preparation and Etherizing Rooms.
E, Toilet.
F, Bath.
G, Employés.
T, Store Room.

I, Officers' Dining Room.
J, Sewing Room.
K, Kitchen.
L, Store Room.
M, Disinfecting Room.

N, Disinfecting Boiler.
O, Ironing.
P, Laundry.
Q, Mortuary.
R, Stairs and Service Lifts.

S, Dispensary Store Room.
U, Boiler Room.
U, Engine and Dynamo Room.

V, Fuel.
W, Basement under Wards.
X, Passenger Elevator (direct acting).

PLATE 32, PAGE 67.

Basement Floor of an Emergency Hospital.

higher and the basement windows lengthened. If the hospital is to be placed on sloping ground, the one end of it may be so high that the basement will be entirely above ground, in which case it may be worth while to lengthen the basement windows and floor, the room thus made to be used as a lecture and drill room for stretcher service, etc.

In all cases the ground floor must be raised at least 18 inches from the ground, and in warm climates and malarious regions it should be at least three feet above the ground, on piers or open arches, and to insure cleanliness the space between piers or arches should be fitted with lattice-work sufficiently close to exclude dogs and fowls.

These hospitals are designed for temperate climates. For cold climates on the northern frontier, and at all posts where the mean temperature of the winter is liable to fall below 20° F., the ward ceiling should be reduced to 12 feet in height, the ceilings in rooms in main building to be correspondingly reduced in height to 11 feet on first story and 10 feet on second story. In such cases the ward windows will be double and 7 feet high by 3 feet wide, the windows in main building to be 6 feet high by 3 feet in width.''

These hospitals are to be heated with steam.

Plate 36, page 75 shows the arrangement for a Regulation hospital of eight or twelve beds, to be constructed of brick and heated by steam.

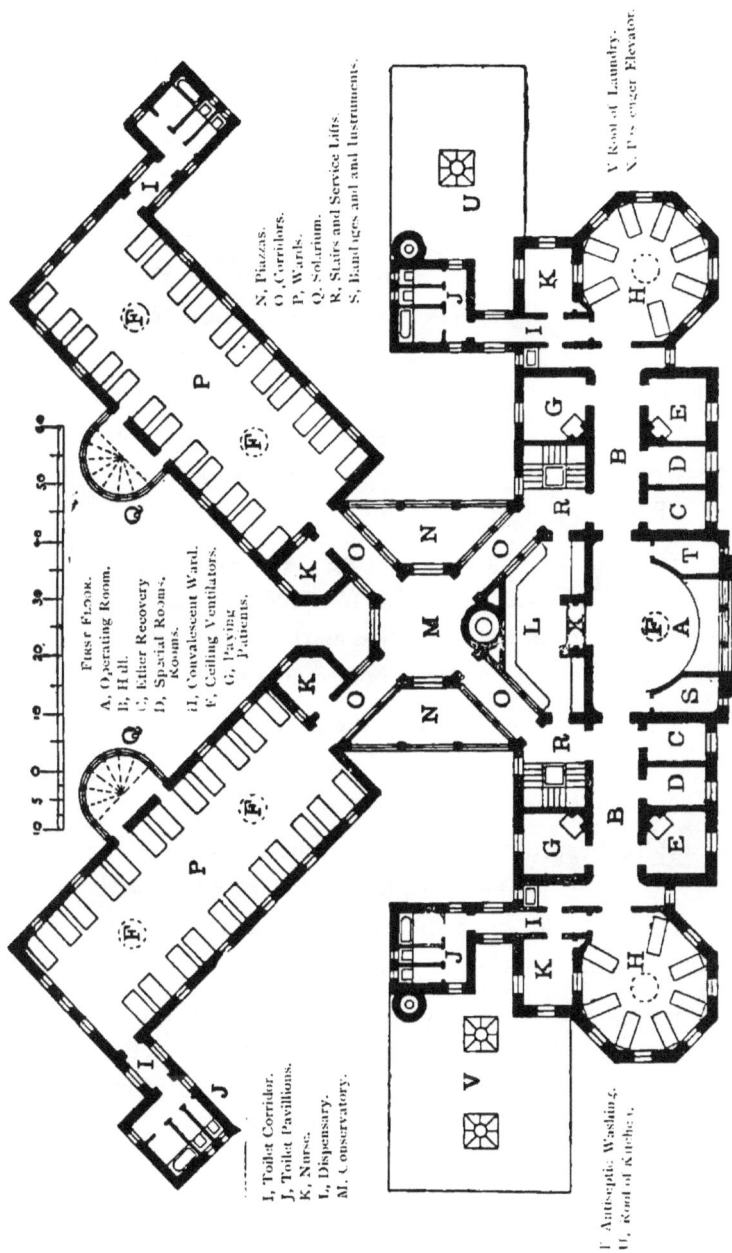

FIRST FLOOR.

A, Operating Room.
B, Hall.
C, Ether Recovery
D, Special Rooms.
  Rooms.
d1, Convalescent Ward.
F, Ceiling Ventilators.
G, Paying
  Patients.

I, Toilet Corridor.
J, Toilet Pavillions.
K, Nurse.
L, Dispensary.
M, Conservatory.

N, Piazzas.
O, Corridors.
P, Wards.
Q, Solarium.
R, Stairs and Service Lifts.
S, Bandages and and Instruments.

T, Antseptic Washing.
U, Roof of Kitchen.

V Roof of Laundry.
X, Passenger Elevator.

PLATE 33, PAGE 69.

First floor of An Emergency Hospital.

The plan as drawn shows a hospital for 8 beds. For 12 beds the ward is simply lengthened in proportion, giving one additional window on each side, the administration building being unchanged.

A, Ward 24' 4" x 30' 1". B, Smoking Room, 9' 2" x 10' 0". C, Toilet Room, 10' 0" x 13' 0". D, Steward's Room, 14' 11" x 17' 2". E, Mess Room, 14' 11" x 15' 5". F, Office, 14' 11" x 17' 2". G, Kitchen, 13' 0" x 14' 11". H, Pantry, 5' 3" x 10' 0". I, Entrance. J, Vestibule, 5' 0" x 13' 3". K, Hall, 6' 0" wide. L, Piazza or covered Veranda, 6' 6" wide.

The ward is one story high.

The administration building is two stories high, and the second story contains four rooms for attendants over and of the same size as the four rooms on the first floor, with a store room over the pantry.

Plate 37, page 77, shows the arrangement of the first floor of a Regulation hospital for 24 beds, to be constructed of brick and heated by steam.

This hospital consists of a central administration building, flanked with two wards as wings, with a detached building in the rear containing kitchen, dining room, isolation ward and attendant's quarters.

Each ward is 14 feet high; though for very cold climates the height may be reduced to 12 feet as above stated. In the 24 bed hospital there is but one set of water closets and bath rooms for the

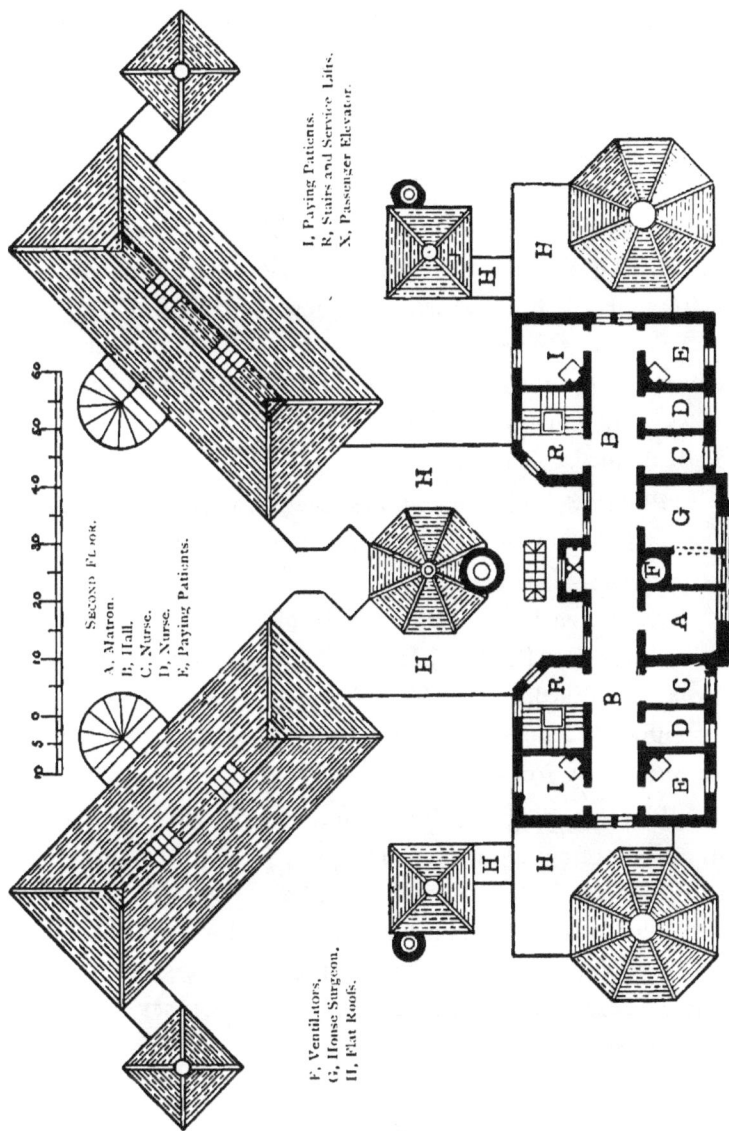

SECOND FLOOR.

A, Matron.
B, Hall.
C, Nurse.
D, Nurse.
E, Paying Patients.

F, Ventilators.
G, House Surgeon.
H, Flat Roofs.

I, Paying Patients.
R, Stairs and Service Lifts.
X, Passenger Elevator.

PLATE 34, PAGE 71.

Second floor of An Emergency Hospital.

whole building. These are in one of the rear rooms of the administration building, as shown on the first floor plan.

A, A, Ward 24' 4" x 42' 1". B, Waiting Room, 15' 10" x 16' 0". C, Toilet Room, 15' 10" x 16' 0". D, Transverse Hall, 7' 6" wide. E, Longi_ tudinal Hall, 6' 0" wide. F, Dispensary, 15' 0" x 16' 0". G, Office, 15' 0" x 16' 0". H, Piazza or covered Veranda, 10' 0" wide. I, Kitchen, 14' 0" x 18' 4". J, Pantry. K, Dining Room, 18' 4" x 23' 0". L, Entrance. M, Porch.

Plate 38, page 79, shows the arrangement of the second floor of this hospital, the divisions being as follows :—

N, Piazza Roof. O, Ward Roof. P, Hall. Q, Store Room. R, R, R, R, Attendants' Rooms. S, Isolation Ward, 14' x 18' 4". T, Bath Room. U, Attendant's Room, 18' 4" x 23' 0". V, Hall opening on outside staircase.

Plate 39, page 81, shows the arrangement of the first floor of a Regulation Hospital for 36 beds, to be constructed of brick and heated by steam.

This hospital is similar to the one for 24 beds, the principal difference consisting in two sets of water closets and bath rooms, giving one set for each ward.

A, A, Wards, 24' 4" x 63' 0". B, Waiting Room, 15' 10" x 16' 0". C, C, Toilet Rooms, 10' 0" x 17' 0".

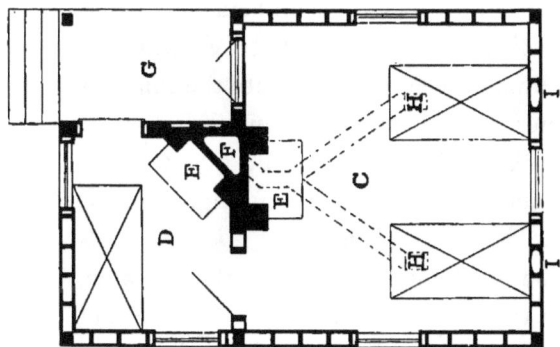

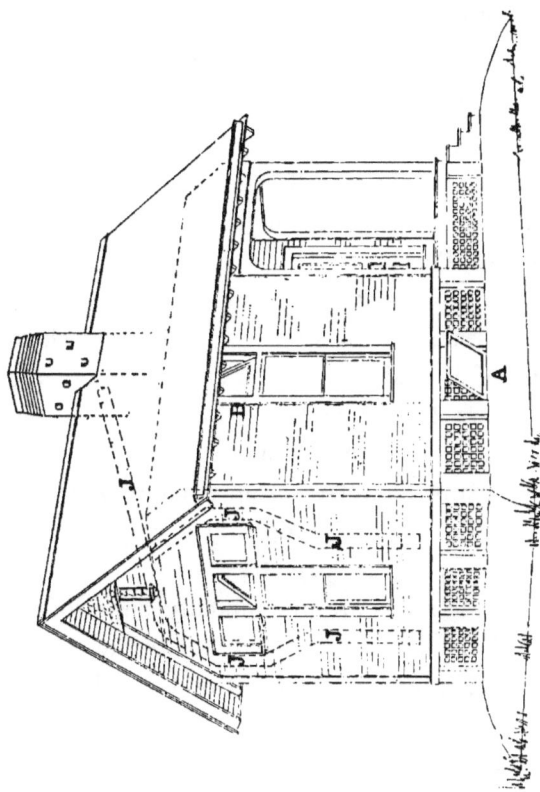

PLATE 35, PAGE 73.

A temporory structure for the worst forms of Contagious Diseases.

D, Transverse Hall, 7′ 6″ wide. E, Longitudinal
Hall, 6′ 0″ wide. F, Dispensary, 15′ 0″ x 16′ 0″.
G, Office, 15′ 0″ x 16′ 0″. H, Piazza or covered
Veranda, 10′ 0″ wide. I, Kitchen, 14′ 0″ x 18′ 4″.
J, Pantry. K, Dining Room, 18′ 4″ x 23′ 0″. L,
Entrance. M, Porch.

The second floor of this hospital is exactly the
same as the second floor of the 24 bed hospital, so
no drawing of it is shown here. Of course the one
story roofs are longer, and it is suggested by the
War Department that the pitched roof attic may be
arranged by mansard construction to contain rooms
similar to those of the second story, as shown by
the elevation in plate 40, page 83.

While these plans are undoubtedly superior to
many that have been suggested heretofore, there
seems some room for improvement in the following
directions :—

1st. The toilet arrangements might be much
improved by placing them in semi-detached wings,
for as the buildings are to be heated by steam the
question of climate need not enter into the consider-
tion. Whether these hospitals are used for surgical or medical cases, or both, it is highly desirable
that complete isolation of plumbing, however
well it may be done, be procured.

2nd. The wards are not sufficiently isolated ;
the air of the wards having more or less close connection with that of the administration building.

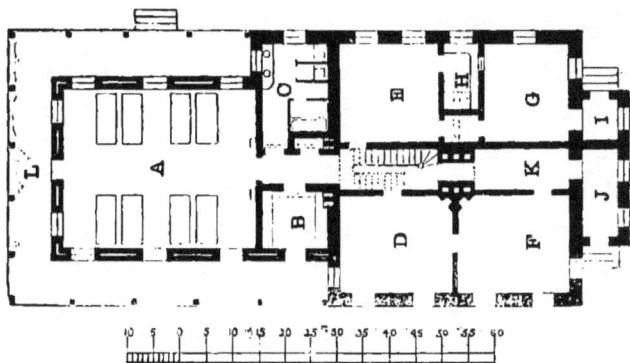

PLATE 36, PAGE 75.
U. S. Military or Post Hospital for eight beds.

3rd. The isolation ward, although accessible only by an outside stair case, is located exactly over the kitchen, where all the food for patients and attendants is prepared, and in summer time, when the windows are open, there must be more or less connection between the air of the two rooms, and air from the isolation ward must in a more or less diluted state enter the kitchen by the windows.

No surer way of spreading disease germs could be well invented. Besides the placing of a patient over a hot kitchen cannot be productive of good to him, no matter how careful the nursing or skilful the medical treatment may be.

As a modification of these plans, that shown in plate 41, page 85, for a 24 bed hospital, is suggested by the writer as perhaps containing less objectionable features.

As a basis for comparison of cost, it may be stated that in plate 37, there are 570 lineal feet of brick wall. In this, 546 feet. In plate 37 there are 445 lineal feet of piazza (measuring in the centre). In this 345 feet (F).

Doors and windows remain nearly the same. The administration department in this building has four corner rooms on both first and second floor.

The toilet rooms are in a semi-detached wing. The piazza on one side of wards may be used for lounging chairs or cots in pleasant weather.

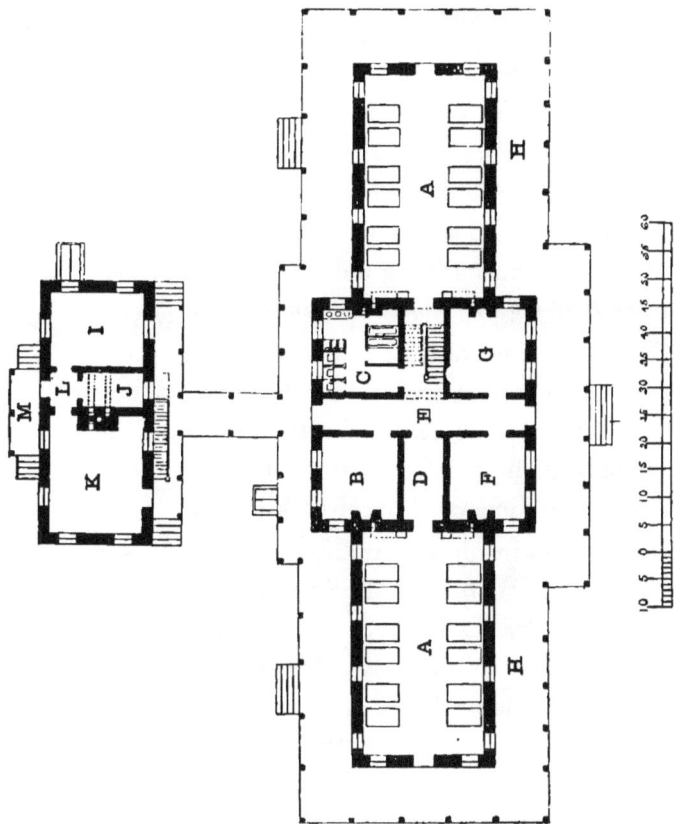

PLATE 37, PAGE 77.
U. S. Military or Post Hospital for twenty-four beds.
First floor plan.

The piazza around the administration building may be reserved for promenading.

The sun-rooms at end of wards may be omitted if desired. The wards are light, having piazzas on one side only. The smell of cooking is kept in the administration building and is fully as isolated, *so far as patients are concerned*, as in plate 37.

The isolation ward does not exist. If one is needed (and there should be one) it should be a frame hut and *really isolated*. In any event, the government cannot afford to set the example of placing an isolation ward over the kitchen.

The letters in plate 41 indicate departments as follows:—A, A, Wards; B, B, Sun Rooms; C, Toilet Pavilion; D, Disconnecting Lobby; E, Hall; F, F, Piazza; G, Dispensary; H, Waiting Room; I, Office; J, Kitchen; K, Pantry and China Closet; L, Dining Room. Proper ventilation can easily be obtained in the Sun Rooms to admit of their being also used as Smoking Rooms without affecting the air of the ward.

Other plans for Hospital stewards' quarters, as recommended by the War Department are interesting, although in the specifications for a building to cost not to exceed $1,200, the clause "The requisite provision to be made for the support of the rear chimney upon the stud partitions below" is unfortunate.

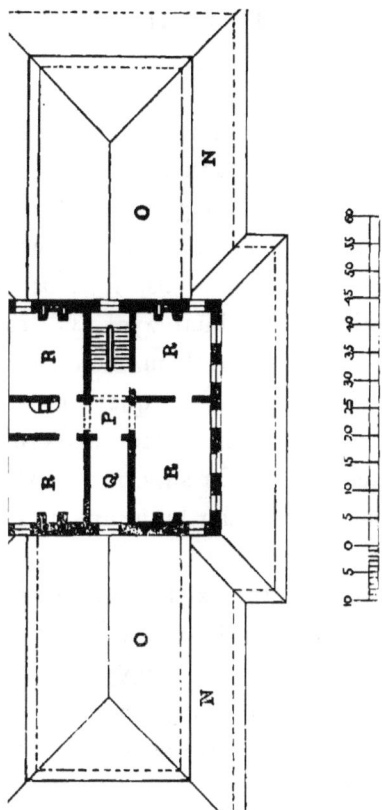

PLATE 38, PAGE 79

U. S. Military or Post Hospital for twenty-four beds.
Second floor plan.

CHAPTER VII.

Much has been written upon the permanency of hospitals, or pavilions connected with them, and probably on this there is a greater difference of opinion than on any other subject. There is no question but that poor materials, in a badly constructed hospital, are productive of evil sanitary results and that a renewal of them should occasionally be made. There is, also, no question but that the quality of much of the better class of work done at the present day is far superior to what was produced in some of the older buildings now being, or about to be, abandoned.

It would seem that the existence of hospitalism at the present day is generally produced by bacteria, (floating in the air or water supply,) and to no subject has so much recent study been given by the medical profession. There can be no question but that in time, and perhaps at no far-distant time, the knowledge of bacteriology will be increased to such an extent that hospitalism, septicaemia and erysipelas will *never* be known in hospitals. When this

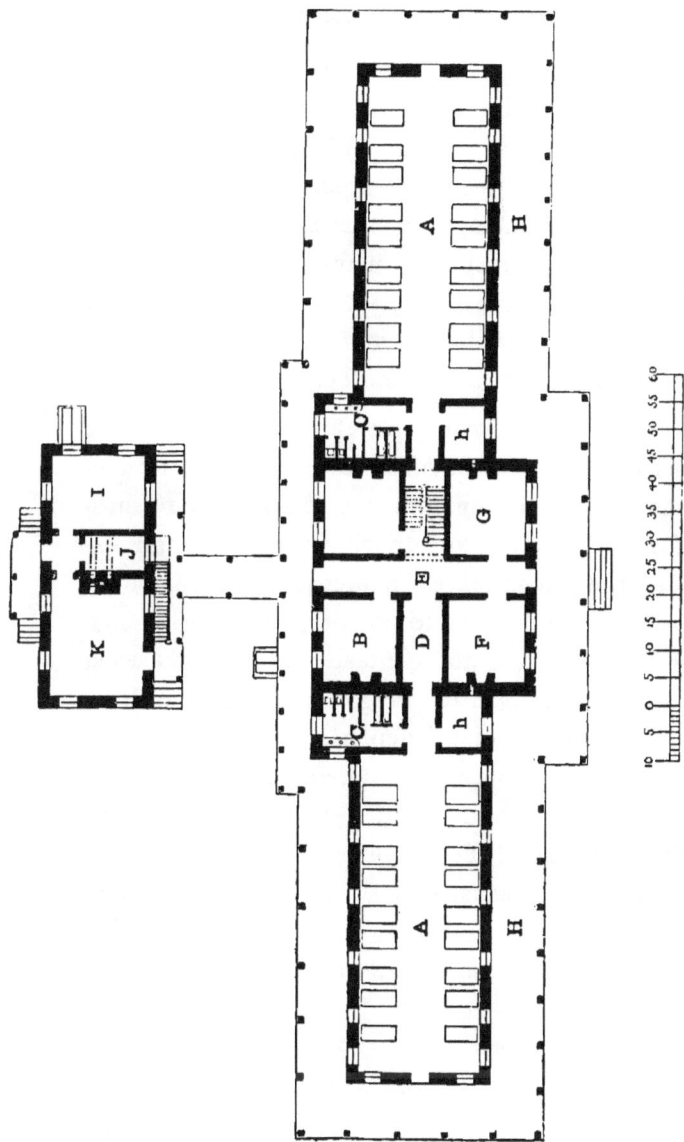

PLATE 39, PAGE 81.

U. S. Military or Post Hospital for thirty six beds.

First floor plan.

state of affairs exists (not because the air is purer, but because the means of resisting or preventing the results of it on patients is discovered), there will no longer be any need of temporary hospitals, and the writer is of the opinion that the success attending the administration of many modern hospitals within the last five years, would seem to warrant that permanent buildings are more desirable than any temporary structures can be.

It is also true that the architect is not expected to perform miracles so often in the construction of permanent as of temporary buildings. Few hospital Boards realize how much things cost, and expect all the conveniences in a $2,500 frame pavilion that they would in a $10,000 or $20,000 brick one. It should be, but is not generally known that wood invariably shrinks after being cut. The use of lumber in our cities and the depletion of our forests is going on at an enormous rate, and it is almost impossible at certain times of the year to obtain dry lumber. Hospital Boards, even when composed of physicians, are usually desirous of speedily providing that protection which they feel is needed by patients, and consequently a temporary building is frequently "run up," and occupied in as short a space of time as possible.

The results cannot be different from those in dwelling houses constructed of frame or wood. In

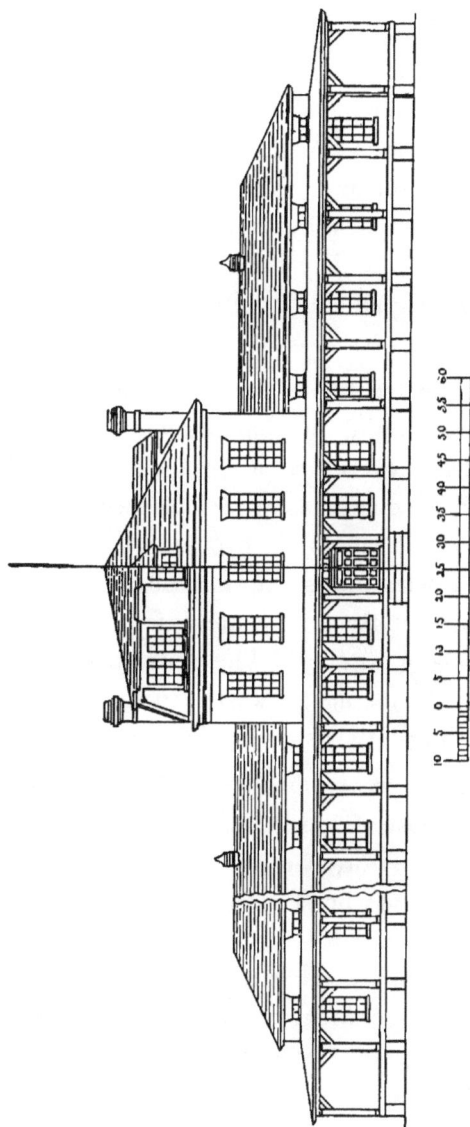

Half elevation on right for 24 bed Hospital.　　　Half elevation on left for 36 bed Hospital.

PLATE 40, PAGE 83.

U. S. Military or Post Hospital.

the course of three to six months cracks appear at
the angles of rooms, and there is a settlement of
the floors where joists are supported by frame work.
This can be prevented only by "crowning" all
joists, and raising them where they rest on the frame
partitions, but while they will then shrink to a level,
they will shrink, and *cracks* must be the *inevitable
result.*

This is not so liable to happen in permanent
as in temporary buildings, but the necessity of having the mechanics return after the building had
been occupied, (sometimes at serious inconvenience,
or hurt, to the patients) is avoided by slower construction.   Such things are not done with such
speed in other countries: we are a fast nation, and
live—and die—in a hurry.   The writer believes that
as frequently "haste makes waste" in the rapid construction of hospitals, as in anything else.

"COMPARISONS ARE ODIOUS."   This is especially true of the cost of hospitals, where the examination made by a visitor, who is is a member of another
hospital board, and who is "looking up" the subject,
will frequently cause him to take home incomplete
or erroneous impressions as to the cost of the building.   In many places it is the custom to state the
cost of a building exclusive of heating, ventilation
and half a dozen other things of minor importance,
but all constituting a comparatively large sum.

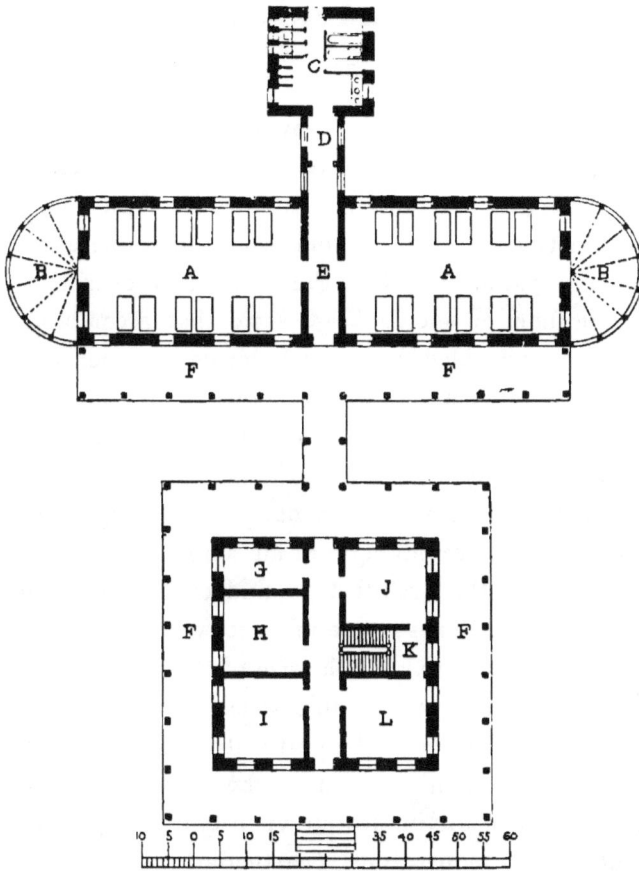

PLATE 41, PAGE 85.

A suggestion for U. S. Military or Post Hospital for 24 beds
First floor plan.

CHAPTER VIII.

OPERATING ROOM.

Most hospitals are provided with an amphitheatre or an operating room. This is necessary in order that the coming generations of physicians may derive experience from witnessing operations on living subjects. It is the custom, however, to arrange them on such a large scale that accommodations are provided for a much larger number than could witness any operation and receive any material benefit from it. The superficial knowledge gained thereby is an illustration of the old proverb —"A little learning is a dangerous thing;" and therefore, before laying out an amphitheatre, it is the duty of the architect to find out the probable number of students whose presence would be allowed at one time, and to keep the size of the room, (allowing, of course, ample cubic space for ventilation), as small as possible. Amphitheatres are necessary, of course, for clinical lectures when a medical school is situated in the immediate vicinity of the hospital. But for many operations, (particularly in abdominal surgery), a large number of observers is not only

Perspective of the McLane Operating Room, Roosevelt Hospital. N. Y.

undesirable but dangerous. With three to seven persons actually engaged in the operation, or assisting at it, an unobstructed view is seldom obtainable by many observers.

Undoubtedly the finest operating room yet visited by the writer is the one in connection with the Roosevelt Hospital, in New York, erected by Dr. Jas. W. McLane and known as the McLane Operating Room.

The exterior is shown on page 87, plate 42, and he arrangement on page 89, plate 43.

The letters indicate divisions as follows:—

A, Examination Room in main building; B, Hall; C, Wardrobes; D, Stack; E, Preparation and Etherizing Room; F, Students' entrance from garden; G, Operating Room; H, Ventilator in ceiling; I, Sky-light (shown on plate 42) in ceiling; J, Brass Rail; K, Room for washing utensils used in operating room; L, Surgeon's Wardrobes; M, Lavatory; N, N, Fifteen Registers in walls near floor. Outside of the rail are benches for students. The floor is marble mosaic; the walls are of white marble to a height of about 8 feet; the wood work and remainder of walls is painted a cream white color; the exposed plumbing pipes are of polished brass, and unlike most exposed work is kept constantly cleaned.

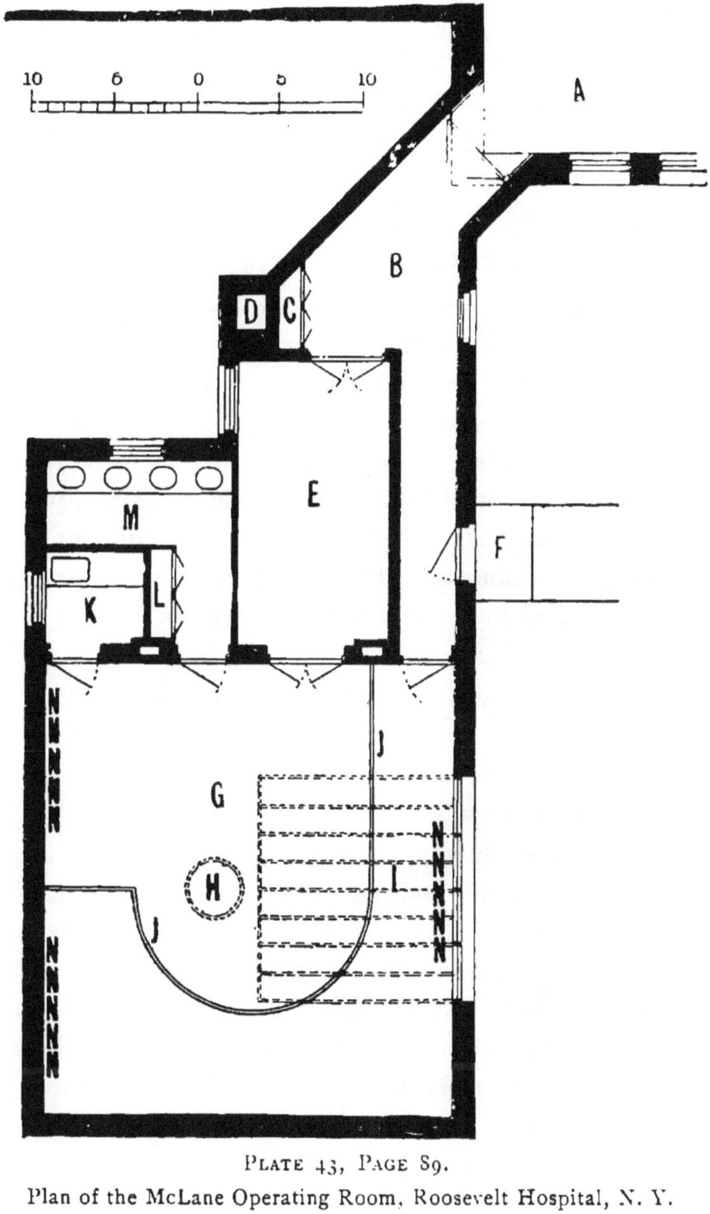

PLATE 43, PAGE 89.

Plan of the McLane Operating Room, Roosevelt Hospital, N. Y.

Indeed, the chief superiority of this room is the cleanliness of it, and the care bestowed on the glass and metal furniture, the instruments submerged in the anti-septic preparations and the abundant supply of fresh warmed air indicated by the writer's anemometer.

If a gallery is desired, no better example can be found than in the operating room of the "Bradlee Ward" of the Massachusetts General Hospital in Boston. Reference is made to the illustrations of it on pages 91, 93, 95, plates 44, 45, 46. The walls are lined with marble to the underside of the gallery, and the floor is of asphalte. A number of small rooms are arranged for the care of instruments and utensils, and rooms for the preparation, or recovery from ether, of patients are provided.

The gallery is not arranged as an amphitheatre, consequently a large number of observers cannot be present at an operation—which is well—but owing to their elevated position a better view can be obtained than by any amphitheatral arrangement.

All operating rooms should be located in separate pavilions, if possible. Nothing is more dangerous than to have the air of the operating room communicating with the administration building or the air from the wards. The light should be north, as nearly as possible, and cross light from a number of sides is not desirable. The height of the room

PLATE 44, PAGE 91.

Perspective of Bradlee Ward and Operating Room.

Massachusetts General Hospital. Boston.

should be sufficient to admit of a sky-light, without increasing the temperature in summer to any appreciable extent; in fact, an artist's studio with the corners rounded is the kind of room desired, and the sky-light and side-light should be *continuous* not separated by a section of the roof.

The operating room should have in connection with it a preparation room, especially in emergency hospitals, which should be in close proximity to the ambulance entrance, and used for the reception of patients, and where they would be examined to ascertain whether their condition would warrant immediate operating or not. From the preparation room, the patients are removed to small rooms, where, if the case is hopeless, they may die without disturbing other patients, or where they may rest a short time preparatory to being operated upon. They are then removed to the etherizing room, then to the operating room, and after the operation to the recovery room; this latter department being especially desirable because many patients in "coming out of ether" are so noisy that others are easily disturbed by them.

The floor of the operating room should be of asphalte, slate or marble in preference to other materials, because these are non-absorbent. The walls should be formed, if the funds will admit, of large slabs of marble, not necessarily white, or slate;

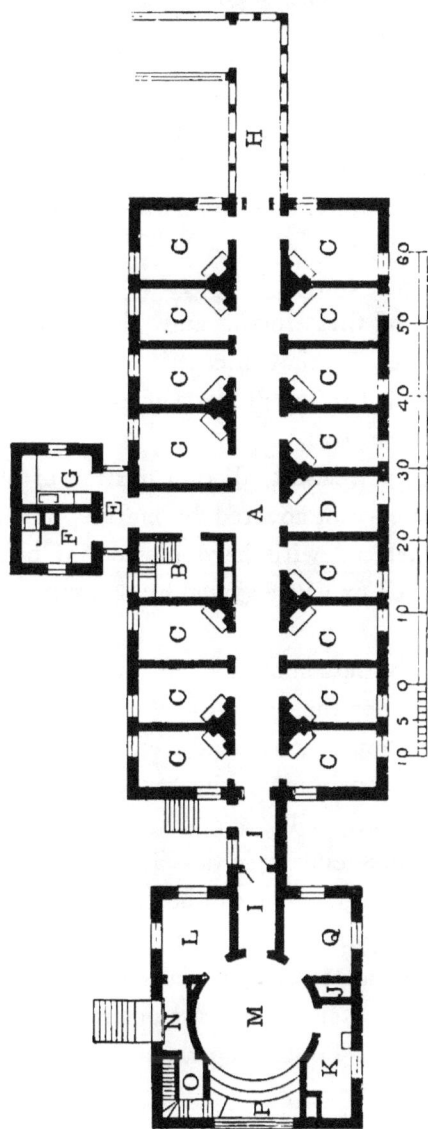

PLATE 45, PAGE 93.

A, Hall; B, Linen; C, Private Rooms; D, Nurse; E, Lobby; F, Water Closet; G, Scullery; H, Closed Corridor connecting with Open Platform; I, Vestibules; J, Ventilator; K, Surgical Attendants; L, Surgeon's Room; M, Operating Room; N, Students' Entrance; O, Stairs to Gallery; P, Seats for Students; Q, Etherizing Room.

where this cannot be done the hardest kind of plastering should be provided. In case it should be deemed impossible to provide such special plastering throughout a hospital, the operating room should be the one place in preference to all others, where it should be used.

Too great care cannot be given to the heating and ventilating of this room, and the windows should be arranged so that in case it were desired, they could be opened in warm weather, *when an operation is not in progress.*

Where funds will admit, all that portion of the walls and partitions not covered by marble, should be of brick or surfaced with hard plaster; if brick is used, light tinted or white glazed brick are preferable to any others.

No drainage outlets should be arranged in connection with or under the operating table or floor of the room, and it should be remembered that unless the ventilation is continuous, whether the room is occupied or not, there will be a stagnation of air which may produce evil results. The supply of fresh air should be as great as can be admitted without causing a draught, in order to give a proper amount for each person to breathe, and to accomplish this best, a number of small inlets is desirable, as is the case in the McLane operating room.

PLATE 46, PAGE 95.

Interior of Bradlee Operating Room. Massachusetts General Hospital. Boston.

CHAPTER IX.

RENEWAL OF AIR IN WARDS.

The frequent repetition of words in this chapter is intended to prevent any misunderstanding.

In connection with the subject of ventilation it may be well to state that in some countries or cities rules have been adopted for the number of cubic feet of air space to be allotted to each bed, or the number of square feet of floor surface each shall occupy. Provided the air is properly warmed on or before its admission to the room, and provided a sufficient amount is exhausted or removed by ventilation, there can be no danger of providing too much air or floor space to the patient.

It should be borne in mind that for surgical cases that are serious, such as the treatment of very bad suppurating wounds and gynaecological cases 3,000 or even 4,000 cubic feet of air space per patient may be desirable, although for ordinary medical cases 1,000 to 2,000 cubic feet of air space is sufficient. Therefore the size of wards depends entirely on the number of patients to be accommodated in

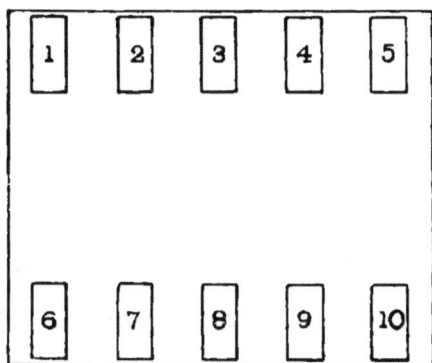

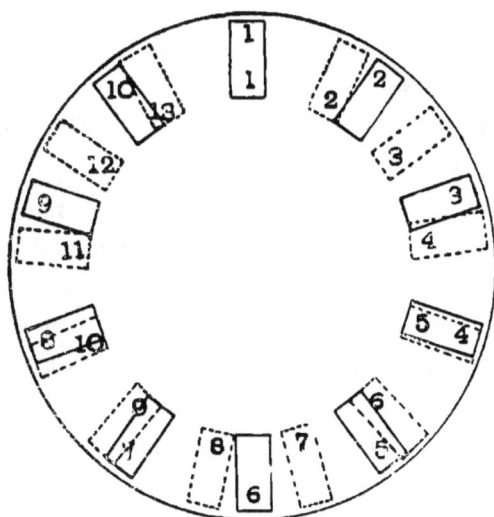

PLATE 47, PAGE 97.
Comparison of Circular and Rectangular Wards of
equal perimeter.

them, and the class of disease with which they are afflicted

The basis of all calculations on the renewal of air in wards must be the patient. If a small amount of air space be allotted him, a more frequent renewal of the air becomes necessary in proportion to the amount of space supplied. The ratio is dependent largely on the floor space and distance between beds. Thus, beds six feet "on centres" in a ward thirty feet wide and twenty feet high should receive a more frequent renewal of air than if the height were reduced to sixteen feet and the beds placed seven and one-half feet on centres, although in both cases the amount of air space per bed would remain the same, viz. 1,800 cubic feet.

The reason for this is that as air moves in a horizontal as well as a vertical direction, and, as some claim, in the direction of the longitudinal axis of a ward, the air breathed by one patient would be re-breathed by his neighbor to a greater extent as the distance between beds decreases.

As one of the objects of hospital construction is to secure fresh air for each patient, close spacing of beds is objectionable. It may be pertinent here to call attention to the fact that in circular wards, as the heads and feet of the patients lie in concentric circles, there is more space between the heads of the beds than between their feet. As a line of given

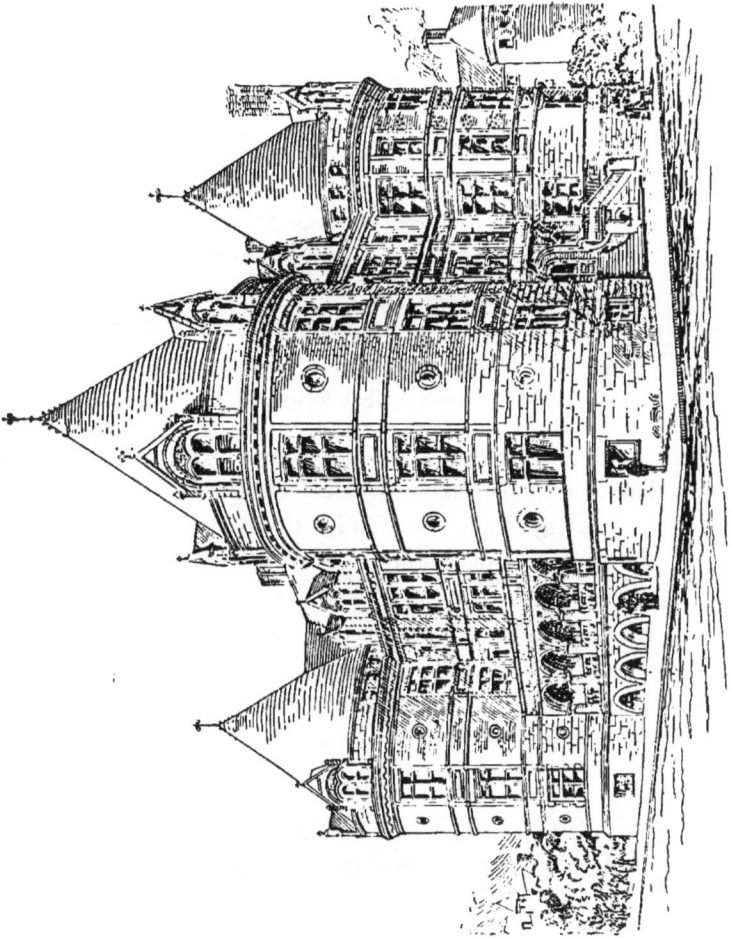

length will enclose more space in the form of a circle, than if it forms a rectangle, it will be readily seen that a maximum amount of horizontal and cubic space is obtained in wards of this shape.

For example, plate 47, page 97, a ward interior 37½ feet long, 30 feet wide, and 16 feet high would accommodate five beds on each side spaced 7½ feet on centres, and contain 18,000 cubic feet of air giving 1,800 feet to each bed. The line enclosing this space would be 135 feet long.

A circular ward, plate 47, page 97, 43 feet in diameter would have a circumference of 135 feet, an area of 1,452 square feet, and if the height be 16 feet, the cubic air space would be 23,232 cubic feet (in comparison with 18,000): and if ten beds are put in there would be 2,323 cubic feet of air space for each (in comparison with 1,800.) But the circular ward could contain 13 beds, giving each 1,787 cubic feet:— only thirteen feet less than in the rectangular ward of the same wall length containing only ten beds.

As regards the spacing of them, the plate shows the circular ward arranged for ten beds in "full" lines and thirteen beds in dotted lines: this gives for thirteen beds a spacing of about ten feet on centres at the head, and seven at the foot, and for ten beds a spacing of about thirteen feet at the head, and nine at the foot. (The consideration of the relative

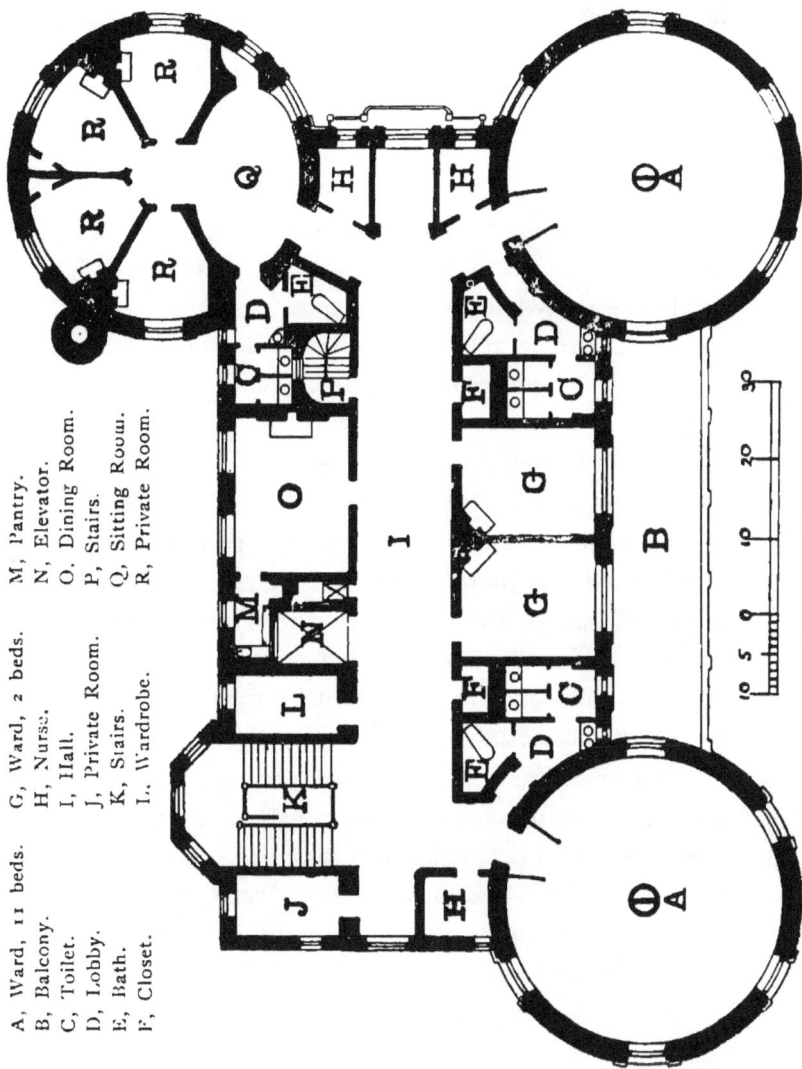

A, Ward, 11 beds. G, Ward, 2 beds. M, Pantry.
B, Balcony. H, Nurse. N, Elevator.
C, Toilet. I, Hall. O. Dining Room.
D, Lobby. J, Private Room. P, Stairs.
E, Bath. K, Stairs. Q, Sitting Room.
F, Closet. L, Wardrobe. R, Private Room.

PLATE 49. PAGE 101.

capacity of rectangular and circular wards is dis-
cussed in chapter X.)

Thus it will be seen that the wall space being
equal, increased head spacing is obtained in circular
wards. The same is true of octangular wards to a
slightly less degree. As the consumption of air by
the lungs is by means of the nose and mouth, it
must be desirable to have the patients lie as far apart
as possible. This aids, by proper ventilation, the
isolation of patients, and decreases the chances of
drafts from in-coming fresh air between beds.

In the New York Hospital, on Fifteenth Street,
arrangements have been made whereby each bed has
an adjustable ventilator under it; a small projection
being built above the floor with the register opening in the side, which serves to a certain extent to
secure an independent ventilation for each bed, and
for this reason helps to isolate each patient.

The amount of cubic air space per capita recommended by writers on Hospital construction varies
from 1,000 to 3,600 cubic feet per bed, the latter
amount being advised for dangerous cases, while in
the Workhouse Infirmaries of England a minimum
of 850 cubic feet is allowed.

It may be considered that any allowance below
1,000 cubic feet is insufficient for severe cases of
any nature. When new buildings are being constantly designed with a view to improving upon past

PLATE 50, PAGE 103.

Perspective of Warren Ward, Massachusetts General Hospital, Boston.

(Showing Solarium removed).

examples there can be no excuse for encroaching on so small an allowance.

The following is a list of some existing conditions in Hospitals of comparatively recent construction.

PAGES.	NAME OF HOSPITAL.	SHAPE OF WARD.	NO. BEDS	CUBIC FT. AIR SPACE PER BED.
57.	City, Boston, Mass.	Rectangular	28	1029
19.	Miller Memorial, London,	Circular	10	1200
56, 57.	New York, N. Y. 15th St.,	Rectangular	21	1800
99, 101.	New York, N. Y. Cancer,	Circular	11	1443
103, 105.	Mass. General, Boston,	Warren (nearly square)	20	2268
107, 109.	Johns Hopkins, Baltimore,	Octagonal	24	1761
111, 113, 115, 117.	Johns Hopkins, Baltimore,	Rectangular	24	1769
77.	New U. S. Military,	Rectangular	12	1194
119, 121.	Royal Infirmary Liverpool.	Circular	18	2052
123.	Royal Infirmary Liverpool.	Rectangular	18	1807

Of course a percentage of the fresh air entering the ward is not utilized, i. e. breathed by the patients, and passes from the inlets more or less directly to the ventilating outlets. The writer suggests that the balance which is breathed be called the effective respiratory energy of the ward. As carbonic anhydride forms the largest ingredient in the impurities of the atmosphere it alone is usually considered as their index in calculations. The amount varies from possibly one part per 10,000 in the country, to perhaps five parts per 10,000 in manufacturing cities where extensive combustion increases the proportion of carbonic anhydride in the atmosphere. The average has been assumed by some authorities at four parts in 10,000, and air containing a no great-

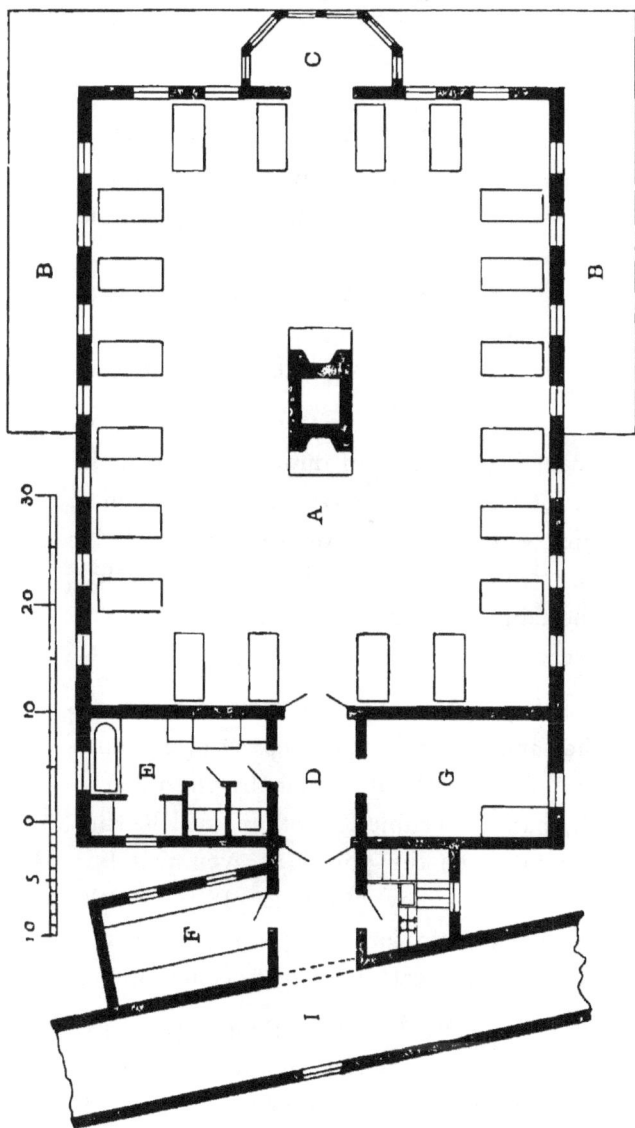

PLATE 51, PAGE 105.

Plan of Warren Ward, Massachusetts General Hospital. Boston.

A, Ward; B, Platform; C, Solarium (not built); D, Hall; E, Toilet Room and Bath; F, Diet Kitchen; G, Nurse; H, Stairs to Basement; I, Connecting Corridor.

er than this proportion may be considered reasonably pure.

The amount of carbonic anhydride produced by an adult in an hour has been proved by careful experiments to be .6 cubic feet, hence if a room contains 6,000 cubic feet he will pollute the air in it in a ratio of .6 to 6,000 i. e. 1 to 10,000. If the air is to be kept at this standard of purity it is necessary to supply 6,000 feet of fresh air per hour for him to breathe.

The normal proportion of impurity existing in the air, viz. four parts per 10,000, should be added to this, which would give the total pollution of the air per hour; from which deductions the writer has prepared the table near the end of this chapter, to which the reader's attention is called.

The limit of allowable total pollution as set down by the same authorities is 10 parts of carbonic anhydride to 10,000 parts of air and by the table it will be seen that 1,000 cubic feet of air must be supplied per hour to keep the purity at even this low standard. While this may do for people in health, it is entirely inadequate for sick people, especially when a number are breathing air in common.

If fresh air is more healthful than impure air, the purer it is the more healthful it must be, therefore the more frequently air is renewed in a ward,

PLATE 52, PAGE 107.
Perspective of Octagonal Ward, Johns Hopkins Hospital, Baltimore.

provided the patient is not in a draft, the more beneficial the result.

In the examples above referred to of a rectangular ward of ten beds and a circular ward of thirteen
beds, the allotment of undivided air space per patient
is 1,800 cubic feet (about), and if we desire to supply
him with 3,000 cubic feet of fresh air in an hour it
will be necessary to renew his allotment $1\frac{2}{3}$ times.

According to the table there would then exist
the constant pollution of 6 parts carbonic acid to
10,000 parts of air. But it has been mentioned before that a percentage of the air supplied is not
breathed by the patient; so to produce a *net* result
of 3,000 cubic feet per hour it will be necessary to
supply a *gross* amount equal to the amount desired
plus the amount lost.

Few tests to discover the percentage of air not
breathed have been recorded, and they are of little value, because the conditions under which they
were made might not,—probably would not, corres-
pond with those of new wards under construction.
The only way to be on the safe side is to provide
a sufficient surplus of air to make up the deficiency,
or to produce the desired result by increasing the
speed of the fans, if mechanical ventilation is used.

The experiments of Dr. Edward Cowles, in the
Accident Ward of the Boston City Hospital, as published by him in a pamphlet in 1879, were very care-

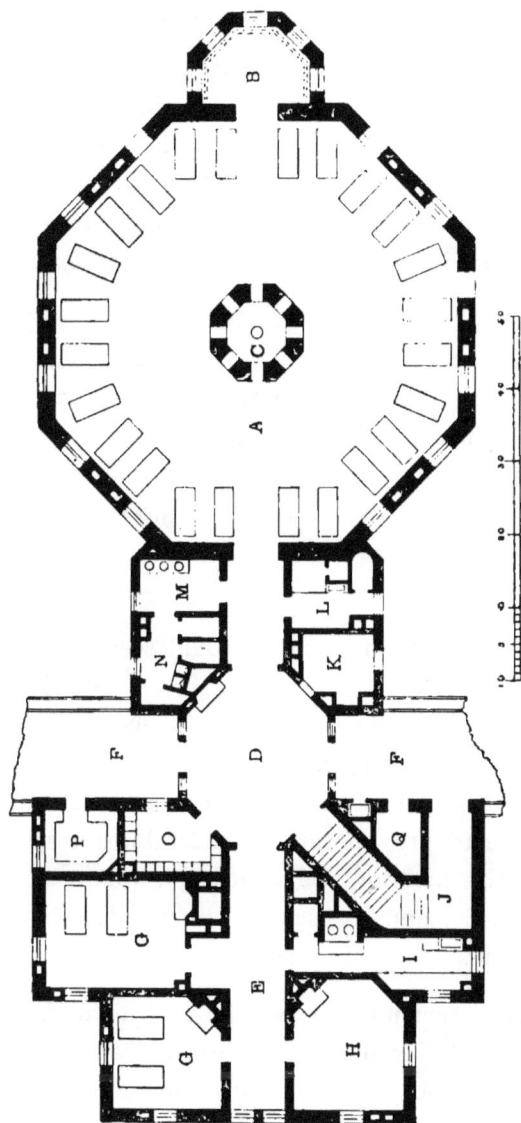

PLATE 53, PAGE 109.

Plan of Octagonal Ward, Johns Hopkins Hospital, Baltimore.

A, Ward; B, Solarium; C, Central Ventilating Chimney; D, Central Hall; E, Hall; F, Open Terrace over Corridor; G, Private Wards; H, Dining Room; I, Kitchen; J, Stairs to Basement: K, Bath Room; L, Nurse's Closet; M, Lavatory; N, Water Closet: O, Linen; P, Patient' Cloth ng: Q, Store Room.

fully made on a thorough and scientific basis; and it was found that the amount of air utilized i. e. "breathed" was about 78 per cent. of the amount supplied by the fresh air ducts. The ventilation was by natural means at the ceiling, though steam coils were placed in the ventilators to aid the exit of foul air.

If we accept this percentage (78) as the average effective respiratory energy of a ward, it will be seen that the amount utilized is not much more than three quarters of the amount received, and therefore to conservatively produce a net result of 3,000 cubic feet we must admit 4,000 cubic feet for each patient per hour.

While conditions may vary, it would seem that it is safe to allow for this percentage of loss. Until the ward is built and the Hospital is in operation, local conditions governing the effective respiratory energy must be largely a matter of conjecture: so it is imperative to provide for a considerable excess of power in the heating and ventilating plant, in order that the demands caused by our variable climate may at all times be supplied.

It should be remembered that the additional air admitted is to make up the amount lost *which is not breathed*, so that the percentage of carbonic anhydride is not increased.

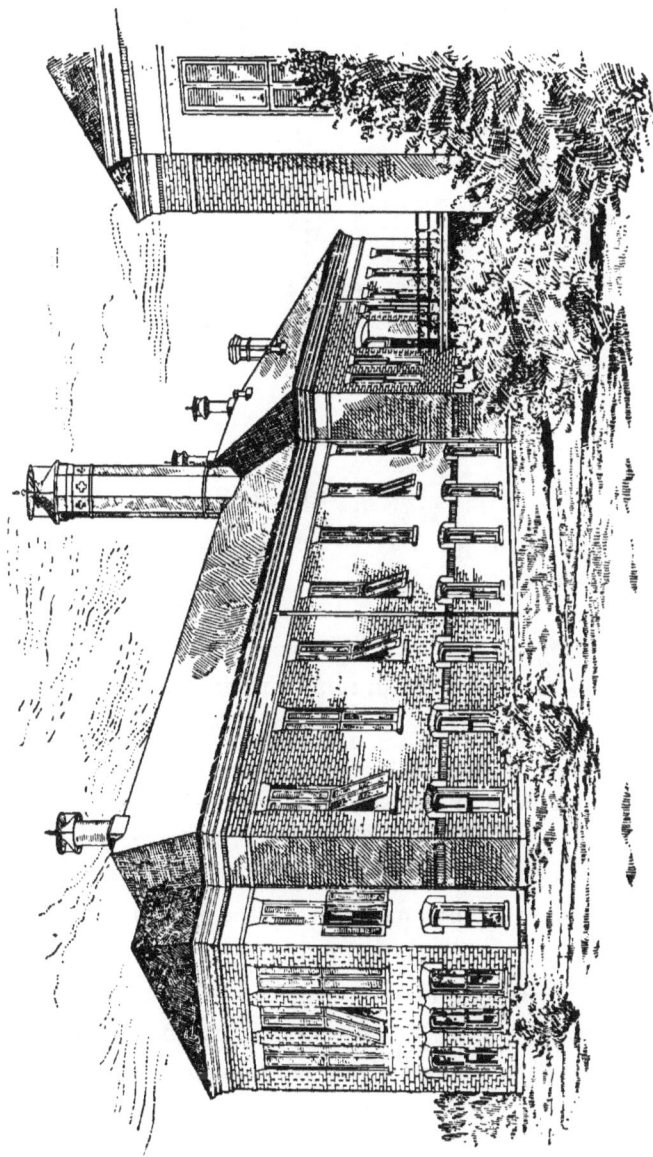

PLATE 54, PAGE 111.

Perspective of Rectangular Ward, Johns Hopkins Hospital, Baltimore.

If the writer's deductions are adopted as a preliminary calculation the following table may be
used.

Gross cubic feet of air supplied per hour for each patient.	Estimated amount of air wasted.	Net cubic feet of air supplied per hour to each patient.	Pollution by Respiration (Parts by volume in 10,000).	Average Pollution in fresh air (Parts by volume in 10000)	Total Pollution, (Parts by volume in 10,000).
4000	1000	3000	2.	4	6
3600	900	2700	2.22	4	6.22
3333	833	2500	2.4	4	6.4
2933	733	2200	2.72	4	6.72
2666	666	2000	3.	4	7
2400	600	1800	3.33	4	7.33
2000	500	1500	4.	4	8
1866	466	1400	4.28	4	8.28
1600	400	1200	5.	4	9
1466	366	1100	5.45	4	9.45
1333	333	1000	6.	4	10

It should be borne in mind that in our example
we are working on a basis of 1800 cubic feet
of air space, and that if more is allotted the patient
it will not be changed so often, and *vice versa*; although the net amount of 3000 cubic feet of supply
should remain unchanged. Therefore the fact that
there is no constant ratio between the amount of air
space and air *supply* should not be forgotten.

For instance, 3000 cubic feet of fresh air per
hour deliverered to a sub-marine diver in his suit
and the same amount delivered in a room to you,
reader, would simply subject him to a greater "draft"

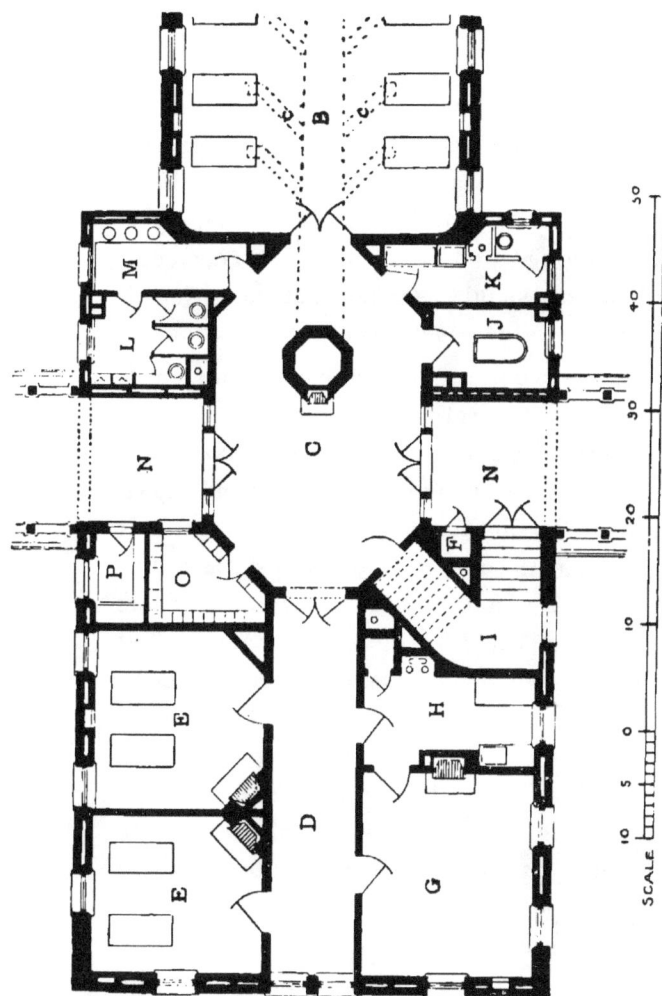

PLATE 55, PAGE 113.

Plan of Head House of Common Ward, Johns Hopkins Hospital, Baltimore.

B, Ventilating Shaft; C, Ventilating Duct: C, Central Hall; D, Hall; E, Private Wards: F, Coal Lift; G, Dining Room; H, Kitchen; I, Stairs to Basement; J, Bath Room; K, Nurse's Closet; L, Water Closet; M, Lavatory; N, Open Terrace over Corridor; O, Linen Room; P, Patients' Clothes.

than you, owing to a more frequent renewal of
the air in the small space he has to breathe in
compared with that which you enjoy. Because he is
confined in a small .space does not enable him to
maintain life with a proportionately less supply of
air.   In arranging for a gross supply of 4000 cubic
feet per hour care should be taken that the patient is
not subjected to injurious drafts.

Returning to our example we have in our wards
10 and 13 beds respectively—assuming that all are occupied we intend to admit and remove 40,000 (gross)
cubic feet of air per hour in the former and 52,000
in the latter.   To do this properly without drafts it
is necessary that the openings for ingress be sufficiently numerous.   In the rectangular ward (doors
being at ends) air could be conveniently admitted at
four points on each side of the ward, between beds.
In the circular ward (13 beds) it could be admitted
at eleven places between beds and leave two places
for door openings.   To admit 40,000 cubic feet per
hour through eight openings would give each 5000
cubic feet, or 83.33 cubic feet per minute and about
1.39 cubic feet per second.   To admit 52,000 cubic
feet per hour through eleven openings would give
each 4727 cubic feet, or 78.78 cubic feet per minute
and about 1.31 cubic feet per second.

Therefore the *velocity* with which the air would
move through a register of one foot (net) area would

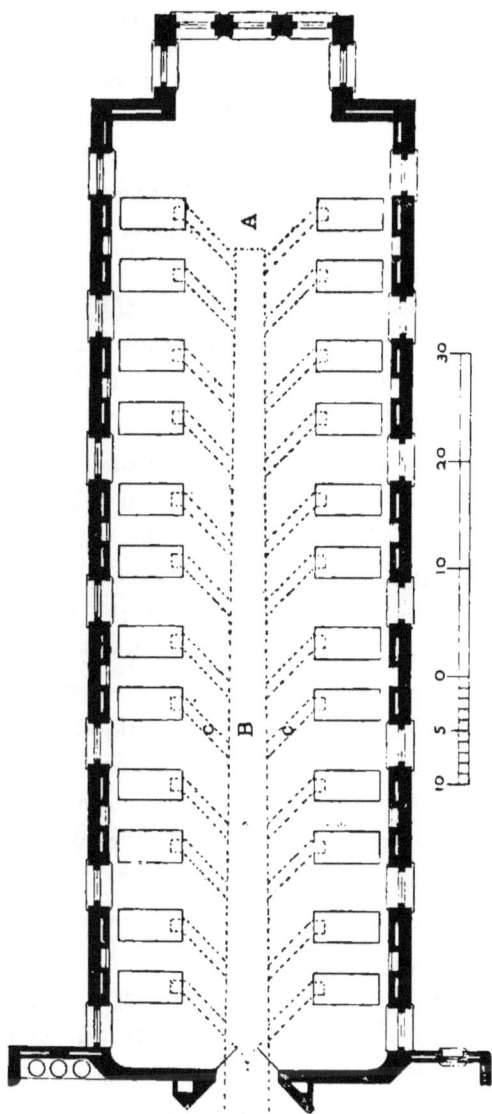

PLATE 56, PAGE 115.

Common Ward of Johns Hopkins Hospital, Baltimore.

A, Ward.   B, Ventilating Shaft.   C, Ventiloting Duct.

be 1.39 *lineal* feet per second in the rectangular ward, and 1.31 lineal feet in the circular ward. If the net area of the opening is increased to two feet the velocity is decreased to one half, and we obtain a rule that the velocity decreases inversely as the area increases and *vice versa*.

The above results for velocity are conservative and within the limits set by Parke in his Manual of Hygiene which gives 5 feet per second as the limit of velocity advisable.

It is desirable to provide a *number* of points of ingress to insure even distribution and to prevent intervening areas of stagnation. If the fan is used for supplying air it should be remembered that the area of the spaces between the bars of the register must equal the area of the pipe supplying the air unless the pipes are much larger than necessary. Valves, if wrought iron pipe, or dampers if galvanized iron or tin pipes are used, should be provided, so that by testing at the various openings a uniformity of velocity may be obtained, when the valve or damper should be *set*.

Where the indirect method of admitting warmed air is used a larger amount of area may be advantageously given. It should not be forgotten that registers, either for ventilating or heating, should never be placed in the floor (unless they are some distance from the beds) as they will surely become

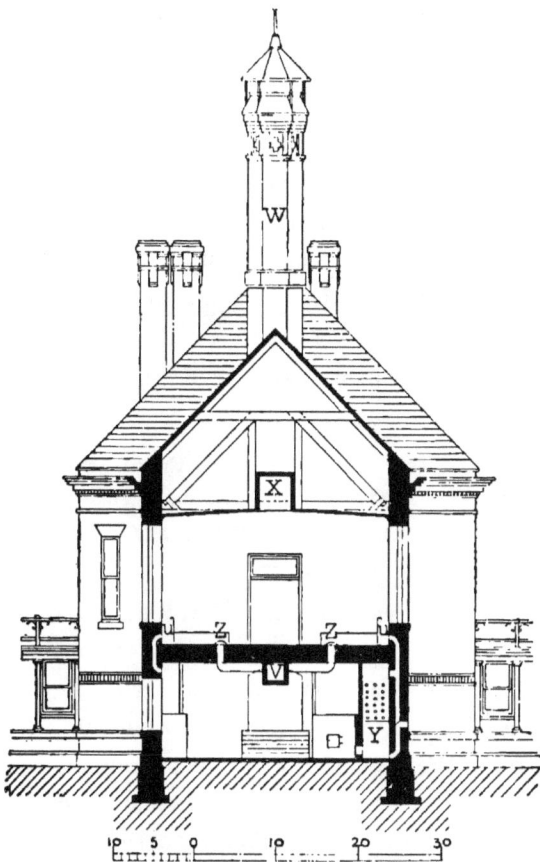

PLATE 57, PAGE 117.

Section of Common Ward, Johns Hopkins Hospital,
Baltimore.

U, Warm Air Registers; V, Foul Air Shaft in Basement; W, Central Ventilating Shaft; X, Foul Air Shaft
in Attic; Y, Heating Coil.

breeding places for germs. All registers, near or between beds, should be vertical, either built into the wall or in the sides of projections built above the floor. The remarks regarding air *space* and air *supply* apply to exhausting the air by a fan as well as supplying it, and care must be taken that the combined areas of exits are equal to those of inlets.

This will admit of the same sized fans being used for both heating and ventilating and running them at the same speed.

RESULTS OF TEST, BY WRITER, OF VENTILATION IN THE McLANE OPERATING ROOM, ROOSEVELT HOSPITAL, NEW YORK, MARCH 13, 1891. *See plates 42, 43 ; Pages 87, 89.*

Cubic space, about 9,100 feet.

Inlets for warmed air (N) are fifteen in number, each register measuring $3\frac{3}{4}''$ x $17''$, or a total gross area of 960 square inches. Allowing a deduction of 25 per cent. for the area (in this case) of the bars of the registers, we have a total net area of 720 square inches, or exactly 5 square feet. The corrected test of the anemometer at each group of registers gave an average incoming velocity of 223 feet per minute, or 1,115 cubic feet of fresh warmed air supplied to the room per minute, or 66,900 per hour. As a fan is used for mechanically delivering a constant supply of warm air by the plenum method, it is evident that the exit of air per hour through the ceiling ventilator must about equal the amount

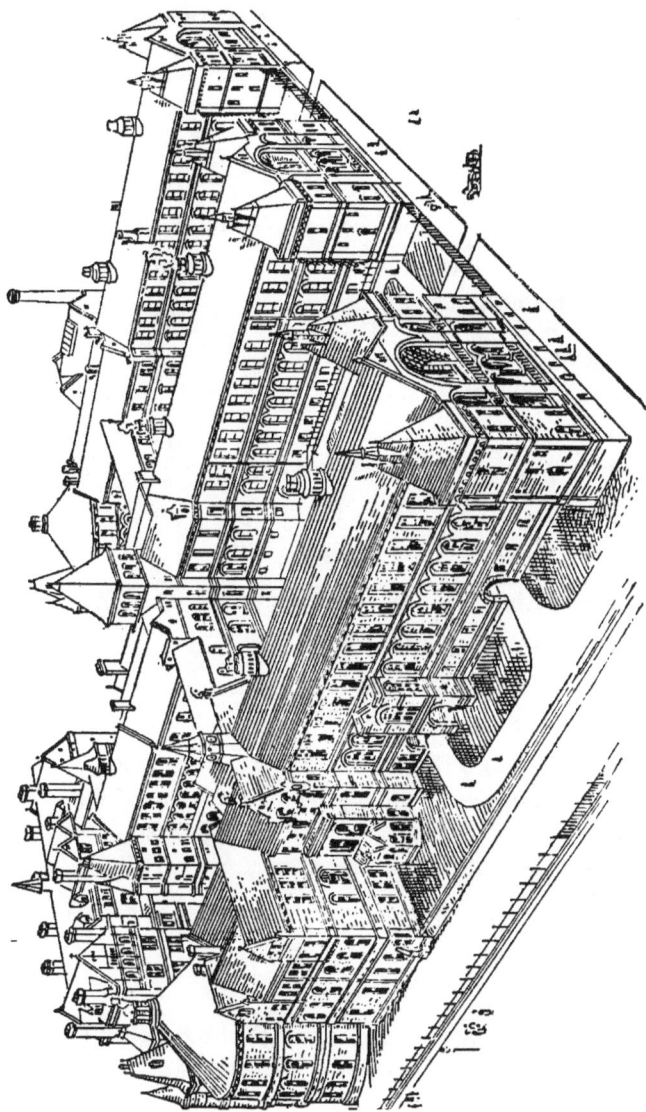

PLATE 58, Page 119.
Birds Eye View of Royal Infirmary, Liverpool.

supplied, although, owing to the height, no accurate reading of the ceiling was obtained. The above is sufficient, however, to show that excellent results are obtained in this operating room, and it may be of interest to state, with reference to future remarks on mechanical ventilation, that the fan is stopped only ten minutes each day for oiling.

RESULTS OF TESTS, BY WRITER, OF VENTILATION IN THE CHILDREN'S WARD AT LAKESIDE HOSPITAL, FEB. 21, 1891. *See Plate 11 ; Page 23.*

The fresh warm air is admitted at 4 registers (B), placed in the floor to save expense in the construction which was kept at minimum.

The vitiated air is removed by four fire-places (D), in the center of the ward.

The four ceiling registers (C), are not depended upon to any extent, although fresh cold air may be admitted through them in winter, and heated air emitted in summer, thus preventing areas of stagnation. All registers may be operated from the floor of the ward. The warm air is passed over steam coils, placed below each floor register, each of which has an independent supply of fresh cold air also controlled from the ward.

The ward contains about 12,530 cubic feet of air, and 12 cribs for children are provided, thus allowing each 1,044 cubic feet: it should be borne in mind

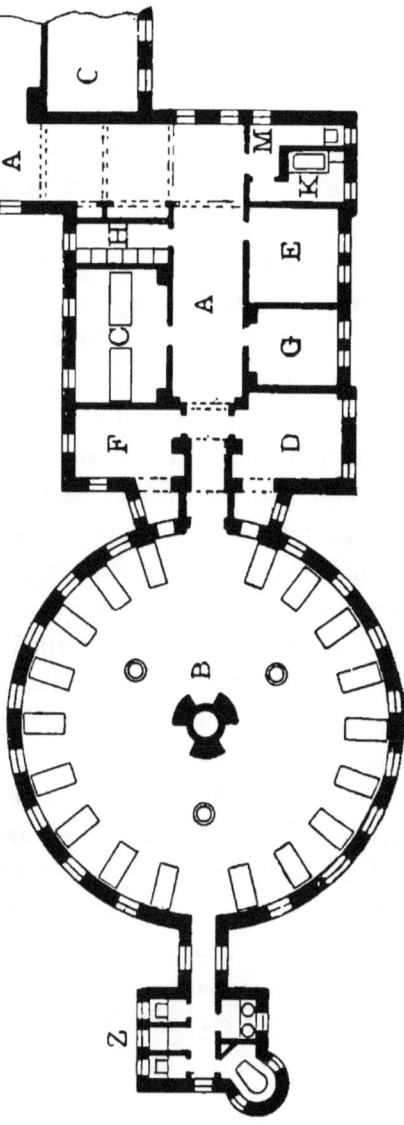

A, Corridors.
B, Medical Ward.
C, Separation Ward.
D, Sister.

E, Doctor.
F, Scullery.
G, Dining and Convalescent Room.
H, Patients' Clothes.

K, Bath Room.
M, Water Closet.
Z, Toilet Pavilion.

PLATE 59, PAGE 121.

Plan of Circular Ward, Royal Infirmary, Liverpool.

that children and old people give off less carbonic anhydride than others. (*Parkes.*)

The size of each fresh warm air register is 14″ x 21″, or a total gross area 1176 square inches and an estimated available net area of 748 square inches. The anemometer showed (testing each) an average corrected incoming velocity of 158 lineal feet per minute or a total supply of 859 cubic feet per minute for the ward.

The total exit area of the four fire-places (at the "throat" of each) is 450 square inches, and the anemometer showed an average corrected outgoing velocity of 158 lineal feet per minute, or a total discharge of 967 cubic feet per minute for the ward.

Owing to the height of the ceiling registers no accurate registration was obtained except that a slight inflow of air was perceptible,—probably sufficient to make up the difference between the warm air supply and discharge.

From the above test it will be seen that the measurable renewal of air in the ward was 967 feet per minute, or 58,020 feet per hour, or an hourly supply of 4,835 feet for each crib, with a constant air space of 1,044 cubic feet each.

The anemometer used by the writer is shown on plate 61, page 125, and is very sensitive.

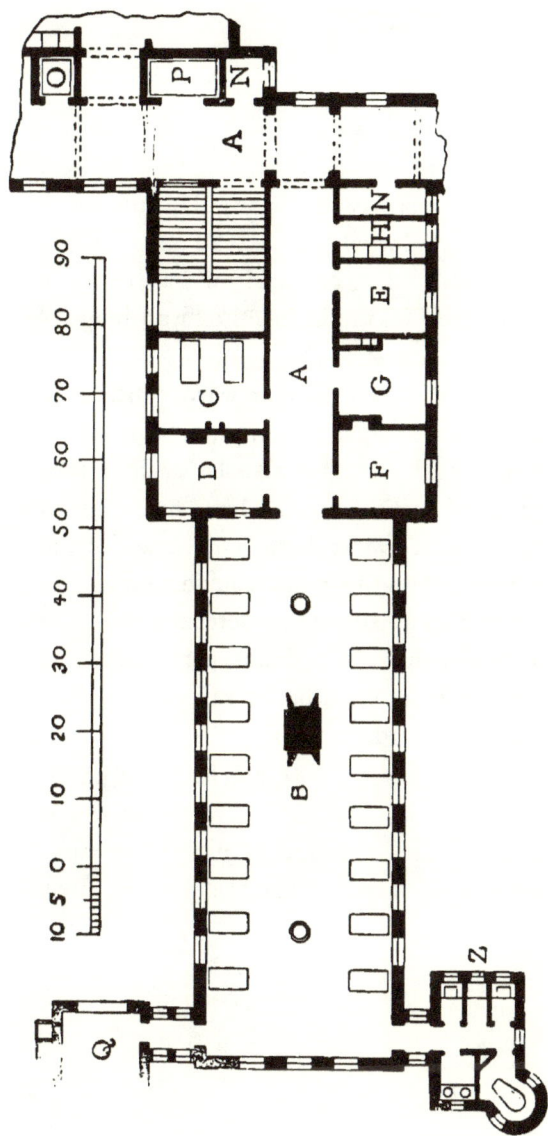

PLATE 60, PAGE 123.

Plan of Small Rectangular Ward, Royal Infirmary, Liverpool.

A, Corridors; B, Ward; C, Separation Ward; D, Sister's Room; E, Doctor's Room; F, Scullery; G, Dining and Convalescent Room; H, Patients' Clothes; N, Housemaids' Closets; O, Service Lift; P, Passenger Lift; Q, Balcony; Z, Bath Room and Water Closets.

CHAPTER X.

RELATIVE CAPACITY OF RECTANGULAR AND CIRCULAR WARDS.

The diversity of opinion regarding the practical value and relative capacity of rectangular and circular wards has been largely caused by an improper basis of comparison. The comparison between the two may be made by taking either the "wall length" or the floor area as a basis.

As it is evident that a circular ward containing the same area as any shaped rectangular ward must have less wall length, it must also be evident that the circular ward will contain less beds, (even though they are placed at both ends and sides of the rectangular ward, though this of course should not be done). It is also apparent, as determined in the preceding chapter, that with equal perimeters the area of a circular ward must be far in excess of any shaped rectangular ward.

Therefore it is important to determine, if possible, when the increased area of a circular ward becomes of no value, or necessitates a useless expenditure of money; for it is evident that there must

PLATE 61, PAGE 125.

Anemometer.

Used for measuring velocity and calculating the volume of air currents.

The lever, A, is an automatic stop.

be a point where the diameter, area and cubic capacity of a circular ward would be largely in excess of all possible requirements, and then it is certain that a waste of money takes place in the construction. That this must be so is evident from a glance at the two diagrams on plate 62, page 127, where the large amount of interior floor space in the circular ward could certainly be better utilized in a ward of rectangular shape.

That this waste space (for it cannot properly be used for any purpose whatsoever,) does not exist in small circular wards may also be seen by referring to plate 63, page 129. The reason for this is that as hospital beds are spaced about eight feet apart on centres, be the ward large or small, the individual floor space for each bed in a circular ward must be a sector, the vertex of which would be at the centre of the ward. In wards of large diameters these sectors must be narrower than in smaller wards (because their chords are nearly alike,) and the distribution of floor area becomes worse as the sector becomes narrower.

The size of the rectangular ward on plate 62, page 127, is 24 feet x 75 feet, perimeter 198 feet, area 1,800 feet, height 16 feet, containing 28,800 cubic feet, or 1,600 cubic feet for each of the 18 beds. The diameter of the circular ward on the same page is 63 feet, circumference 198 feet, area 3,117 feet,

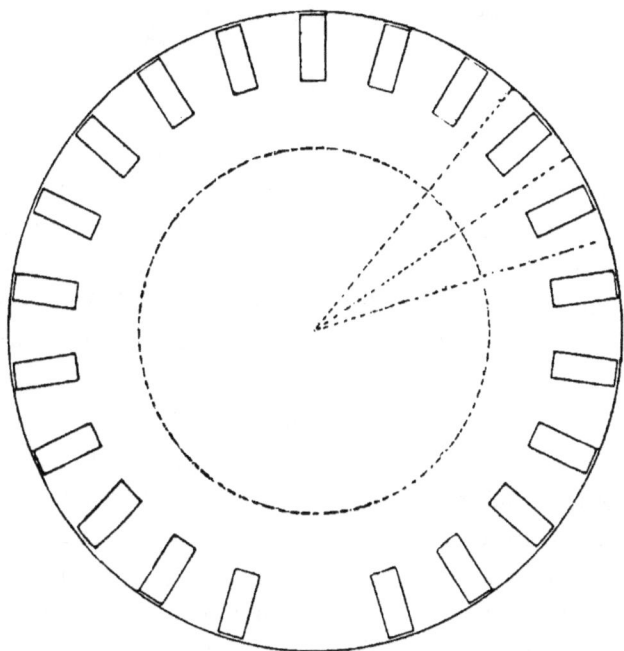

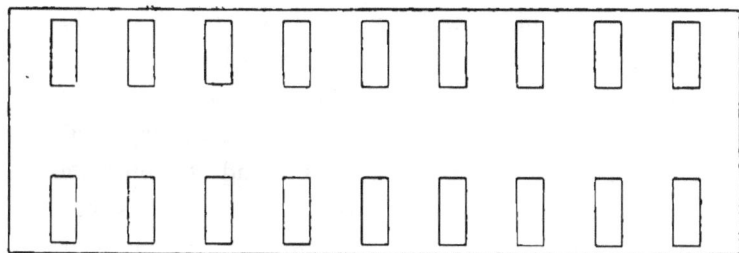

PLATE 62, PAGE 127.

Relative Capacity of Circular and Rectangular Wards of equal perimeter.

height 16 feet, containing 49,872 cubic feet, or about 2,375 cubic feet for each of the 21 beds.

The beds in each of these wards are drawn eight feet on centres: therefore it is evident that quite a large circle could be removed from the centre of the circular ward and leave a space between the concentric circles, which would give each bed 100 square feet, or the same area as in the rectangular ward.

The area of the circular ward is 3,117 square feet, and the dotted circle has a diameter of nearly 36 feet and contains 1,017 square feet, leaving a space between the circles of 2,100 square feet for 21 beds, or 100 square feet each. This space cannot be used for stairs, for a direct communication would be made with the ward above; it cannot be utilized for a central fire place and ventilating stack, for it is too large, and several on its circumference would obstruct the ward. It is much larger than is necessary for keeping bottles, dishes and other articles used in the ward, as is the case at the Antwerp Civil Hospital (see pages 12 and 15), and it would be criminal to make use of it for a day room or sleeping room for a nurse. Therefore it will be seen that a large circular ward wastes much valuable ground.

Let us now refer to plate 63, page 129, and we have a rectangular ward 24 feet x 43½ feet, perimi-

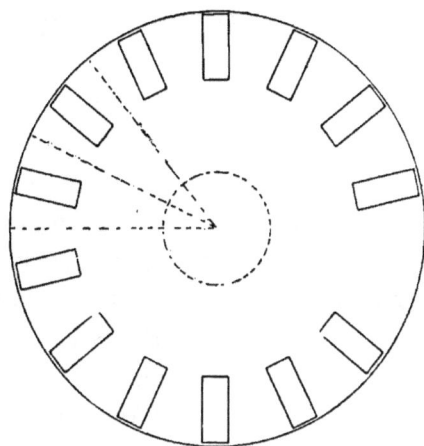

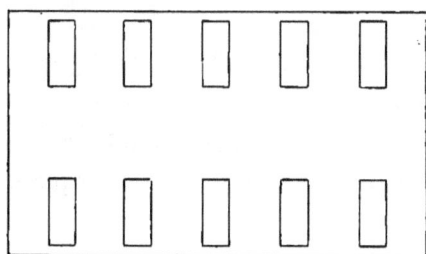

PLATE 63, PAGE 129.
Relative Capacity of Circular and Rectangular
Wards of equal perimeter.

ter 135 feet, area 1,044 feet, height 16 feet, containing 16,704 cubic feet, or about 1,670 cubic feet for each of the 10 beds.

The diameter of the circular ward, on the same page, is 43 feet, circumference 135 feet, area 1,452 feet, height 16 feet, containing 23,232 cubic feet, or about 1,787 cubic feet for each of the 13 beds.

Proceeding, as before, to remove a central circle so as to leave the same area for each bed. as in the rectangular ward (in this case 104.4 feet), we find the diameter of the inner circle is only 11 feet with an area of 95 feet.

This small space *can* be used, and to advantage, by building a brick stack with four fireplaces and four ventilating flues, which would take up, say, 65 square feet, leaving only about 30 square feet of surface wasted.

But, in this case, *is* it wasted?

By referring to both plates it will be noticed how much nearer the so-called waste space is to the bed in the small ward, and the greater divergence of the radiating lines bounding it. It will also be seen that with the same perimeter (135 feet) the circular ward contains three more beds spaced the same—at the centres—than the rectangular ward under the same conditions.

Thus it will be seen that a large circular ward compared with a large rectangular ward, wastes

10 5 0   10   20   30   40   50   60

A

C   C   B

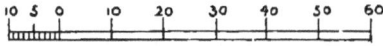

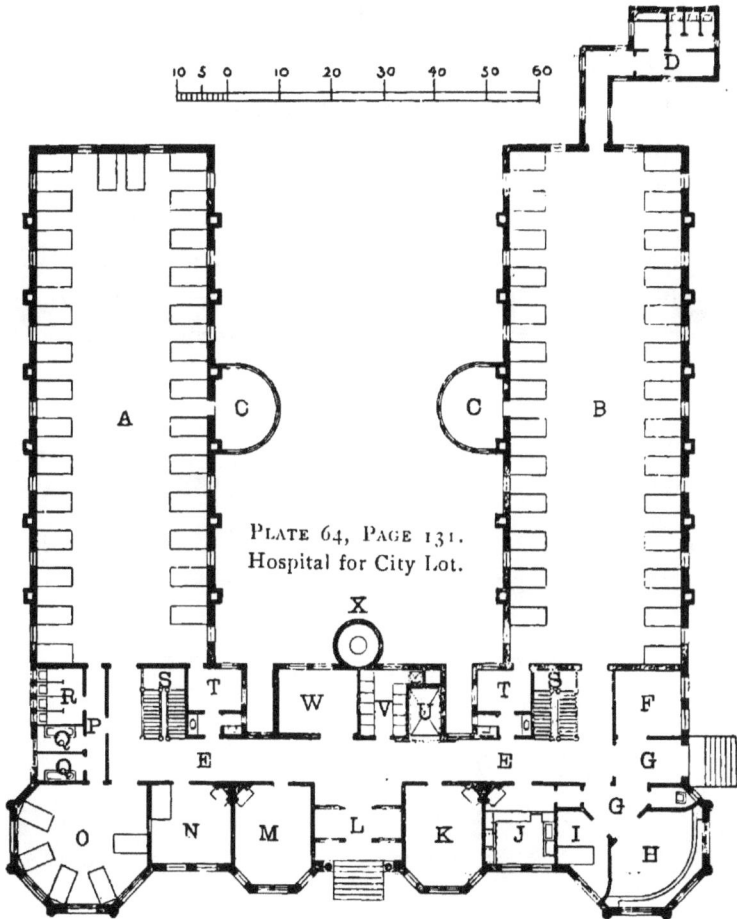

PLATE 64, PAGE 131.
Hospital for City Lot.

X

R   S   T   W   V U   T   S   F

Q P   G

Q

E   E   G

O   N   M   L   K   J   I   H

A, Medical Ward, men.
B, Surgical Ward, men.
C, Solarium.
D, Toilet Pavilion.
E, Hall.
F, Etherizing Room.
G, Ambulance Lobby.
H, Operating Room.
I, Recovery Room.
J, Dispensary.
K, Surgeons' Room.
L, Vestibule.

M, Office.
N, Matron.
O, Special Ward.
P, Toilet Passage.
Q, Bath Rooms.
R, Water Closets.
S, Stairs.
T, Nurse.
U, Elevator.
V, Linen.
W, Reception Room.
X, Ventilating and Smoke Stack.

Officers' and Attendants' Rooms. Kitchen, etc., in third story.
Boilers, Laundry, etc. in front basement.

room, but that (as stated on page 10) circular wards,
of from 30 to 45 feet diameter, are desirable, and
contain greater capacity, under more advantageous
conditions, than small rectangular wards.

The writer has endeavored to consider the subject in its true light, on a proper basis of comparison to offset the results of the misguided criticism
of *all* circular wards as advanced in a recent English
publication.

CHAPTER XI.

HEATING.

Various systems of heating hospitals have been devised; some of them are successful on account of simplicity of construction; others are successful in operation when the direction of the wind is favorable, and others seldom work well under any conditions. It may be stated that, as a rule, the simpler the method of admitting heat to wards, the more equable will be the result obtained. It should be understood that to *successfully* heat a ward it is necessary to remove an amount of air equal to the quantity of fresh air admitted, because where this is not the case, there must be more or less stagnation or lack of circulation in various parts of the ward.

Heating may be divided into several main classes.

First, by reflection from an incandescent surface.

Second, by direct radiation, so-called, or by means of radiators, for steam or hot water, set in the ward.

Third, by indirect radiation or the admittance of air through such radiators, before passing it into the ward.

Fourth, by forcing fresh air, by means of a fan, through a large coil of pipes for steam or hot water, and thence conducting the air by means of pipes to various portions of the ward and admitting it through registers in the floor or walls.

Of all these methods of heating the first is the most primitive and unscientific, although there is no question but that the open fireplace is the best natural *ventilator* for removing heavier gases which accumulate near the floor. But a ward fire-place should not be depended upon for giving out heat; in fact, fire-places in hospitals might be constructed without the usual projecting back for throwing out heat by reflection so as to prevent overheating the ward at certain points; but a brisk fire should be burning at all times where such heat is not objectionable, for the sole purpose of creating a strong upward draft and removing foul air from the ward. The heating of the ward should be obtained by one of the three latter ways described; of them the least desirable is the use of the direct radiator or coil standing in the ward. While this is permissible in an administration building, and affords a very ample means of heating where the officers, *who are well*, pass much of their time, it would not be a satisfactory means of warming the air breathed by patients. As there is no direct supply of fresh air from the outside to these radiators, except by the

PLATE 65, PAGE 135.

Perspective of Hospital for a City Lot.

windows, and such places where air may leak
through walls or pass through doors, there is some
probability of the air becoming stagnant and dead.
The frequent packing of valves, and the frequent
opening of cocks to release the confined air in the
radiator and to produce a circulation of steam, cre-
ates an almost constant hissing sound, very disa-
greeable even to well persons and more so to *patients*
of a nervous temperament.

The third method of admitting air by means of
openings connected with each separate coil, is on
many accounts the best; by it no one point is depended upon for supplying all the fresh air for the
building, and each ward or room may be controlled
by the nurse or patient from the ward, so that the
air may be admitted to any extent desired. This
should be accomplished by extending the valves
through the floor of the ward, and by means of a
damper for the admission of cold air from the outside. The damper should not be arranged to work
with chains, unless it is unavoidable, on account of
the rattling which they cause and the liability to
become broken or slip from the pulleys over which
they pass. Much has been said about the location
of the fresh air inlets; some writers claiming that
the air should be taken from near the roof. It
should be remembered, however, that in cities the
conductors which convey the rain water from the

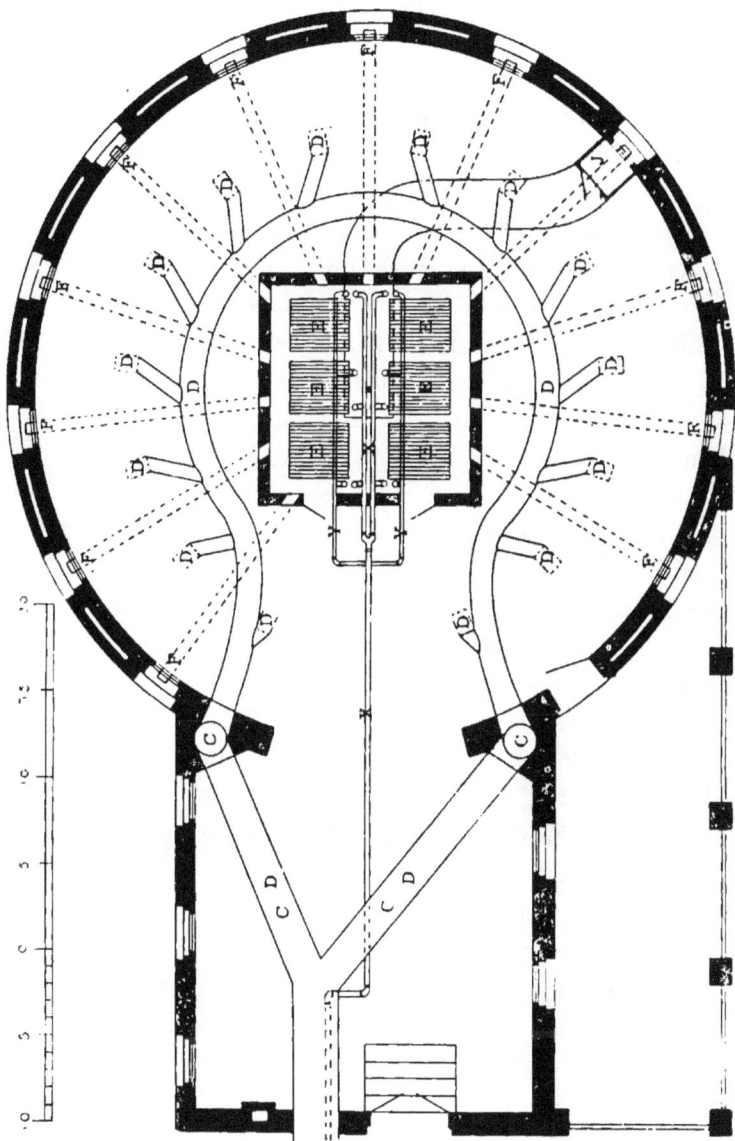

PLATE 66, PAGE 137. Basement Plan of a Suggestion for a Circular Ward.

C, Ventilating Stacks; C, D, Main Ventilating Shaft; D, Ventilating Duct; E, Heating Coils; F, Warm Air Registers; J, Cold Air Shaft; X, Supply Steam Pipes; Y, Condensed Steam Pipes.

gutters are usually connected with the sewers, and
that there is frequently a stream of sewer gas exud-
ing from their openings at the gutters, which
might contaminate the fresh air admitted to the
patients, were it obtained from near these points.
Therefore the writer is of the opinion that it is far
preferable to admit the fresh air quite close to the
ground, taking care that ashes and accumulations
of dirt are not allowed near these openings and that
if possible, grass be grown for a considerable distance around them.   It is far preferable to take the
chances of dust being admitted to the ward and in-
haled by the patient than the probability of admitting air which has been poisoned by the germs and
baccilli which sewer gas may contain.   In connection with this it would also be well to state that as it
has been, and still is, the custom in the erection of
buildings to put in area *cesspools* of iron, which are
connected with a drain, for the removal of rain
which may fall into the areas, they should be omitted in all hospital buildings, because if fresh air is
admitted from near the ground, these area cesspools
after a continuance of dry weather (which would
evaporate the water from the trap,) would admit
sewer gas as described above.   Another reason why
this method of admitting air at various points to
the individual coils is preferable to any other, is because with the fan ventilation a given quantity of foul
air is removed from the ward and the tendency is to

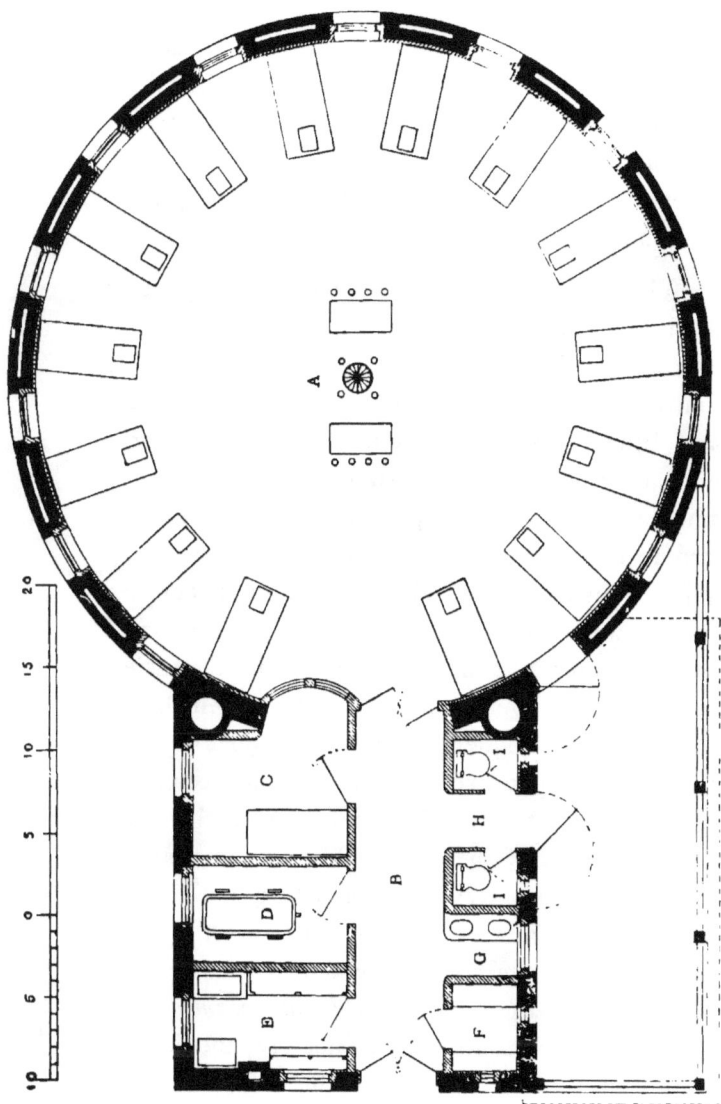

PLATE 67, PAGE 139.   First Floor Plan of a Suggestion for a Circular Ward.

A, Ward; B, Hall; C, Nurses' Room, D, Bath Room; E, Diet Kitchen; F, Linen Closet; G, Lavatory; H, Passage; I, Water Closet.

cause a vacuum ; then fresh air will admit itself to
readjust the equilibrium and there can be little op-
portunity for a draft because of the numerous
divisions through which the fresh air is admitted,
although the exhaust fan might be removing many
thousand cubic feet of air per minute.

The fourth method, while advocated by some
as being preferable to all other methods, may be used
in two ways, first, by forcing in warm fresh air and
admitting it through registers to wards, and at the
same time removing the foul air by means of the
ventilating fan; or by admitting the warm air in the
same way and depending upon the pressure of it in the
ward to force out foul air through the ventilating
shaft. This latter method should never be adopted
for the simple reason that the foul air will be forced
out of the openings through which it has the easiest
and most direct communication with the ventilating
stack, leaving, frequently, the lower openings for
ventilating inoperative and causing the accumula-
ting of stagnant air and carbonic anhydride over
the lower levels of the ward. The disadvantage of
blowing in fresh air and exhausting foul air by
means of two fans is open to criticism—the criticism that *where a direct method will work satisfact-
orily a mechanical devise should not be employed;*
it is where a direct method will not give the desired
result that the fan is most valuable. This method

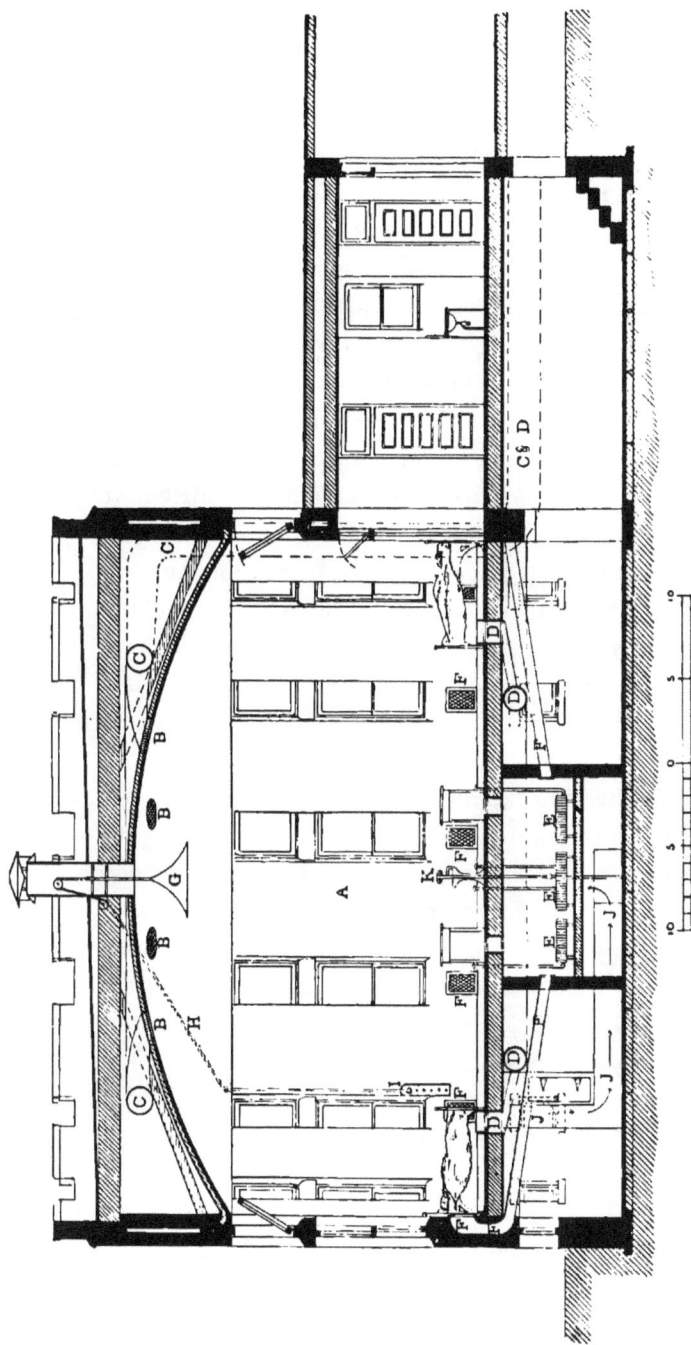

Section of A, suggestion for a Circular Ward.   PLATE 68, PAGE 141.

A, Ward; B, Ventilating Registers in Ceiling; C, Ventilating Shaft Under Roof; D, Ventilating Ducts and Shafts in Basement; E, Steam Coils; F, Warm Air Pipes and Registers; G, Ventilating Damper; H, Chain to Damper; I, Damper Regulator; J, Cold Air Shaft; K, Damper to Cold Air Shaft; C & D, Main Ventilating Shaft.

of heating by forcing in warm air with a fan places the entire hospital at the mercy of any malicious person who might be disposed to poison the entering air which would be admitted by means of the various ramifications of pipes to all the different rooms. Again, if any disease germs or odors from any manufacturing establishment or the smell of cooking, were strong in the vicinity of the fresh air source, it would also be disseminated in a mild form throughout all the wards. It is also a fact that this method of forcing warm air into the wards, when awarded to the "lowest bidder," is an incentive to him to place a much smaller number of inlets than should be made, thereby increasing the velocity; while this is pardonable in factories it should never be allowed in the heating of hospitals, where drafts of warm air are as injurious as cold. But where funds will permit, and a constant certain supply of fresh air can be obtained at one point, the use of a heating fan and a ventilating fan prevents the direction of the wind from forming a varying factor in the warm air supply. Having then decided that we would prefer warm fresh air admitted naturally at numerous points, it remains to be decided whether steam heat or hot water heat shall be used. There is a decided advantage in favor of hot water, on some accounts, the most important of which is the low temperature which the water may be main-

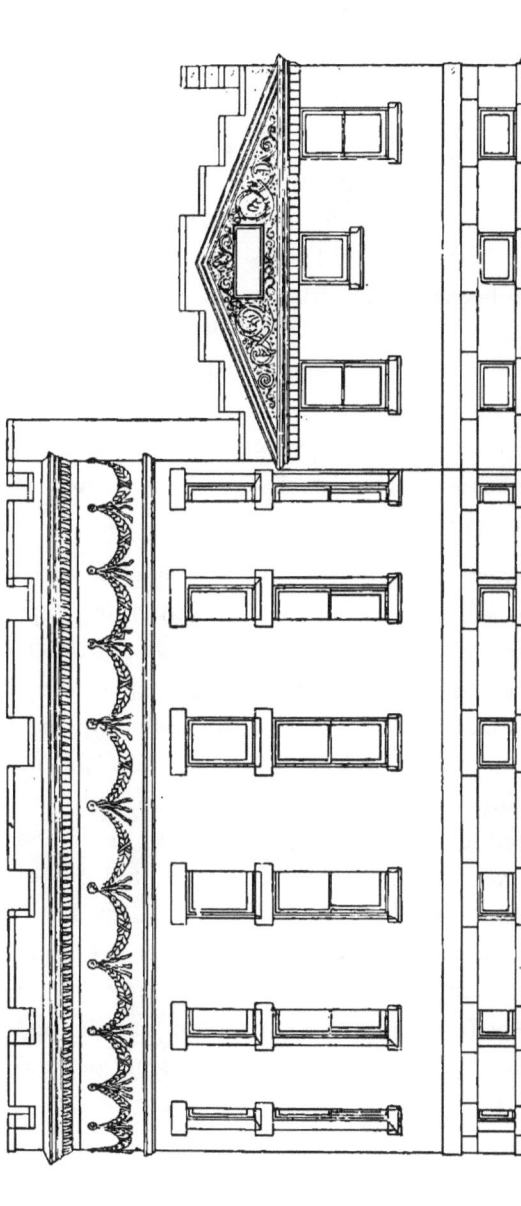

PLATE 69, PAGE 143   Side Elevation of A Suggestion for a Circular Ward.

tained at in temperate weather, thereby giving a
more temperate heat, while in cold weather the heat
of the water may be raised so that the temperature
of the ward may be maintained alike at all times.
This cannot be so well done with steam, or rather
it would be better to say this is not so easily done
with steam because committes desire that the work
cost as little as possible, and unless it is "so nomi-
nated in the bond" the steam fitter will not sub-
divide his coils to admit of operating one or all sections according to the temperature of the outside
air. When this is done steam can be governed as
easily as hot water.

The most important part in either of these
methods of heating is to see that the coils are kept
clean. The writer has seen in almost all instances,
where these methods have been used, great accumulations of dirt and filth drawn in from the outside air and covering the coils which warm the air
before passing it to the ward. As these coils are
usually enclosed in galvanized iron, or tin lined
wooden boxes, they are "out of sight and out of
mind;" therefore, while great care is taken in keeping floors and bedding clean, in order that the patients may have fresh air to breathe, the fresh (?)
air which is being admitted is passing over a vast
area of filth, and bearing with it a floating population and dust.

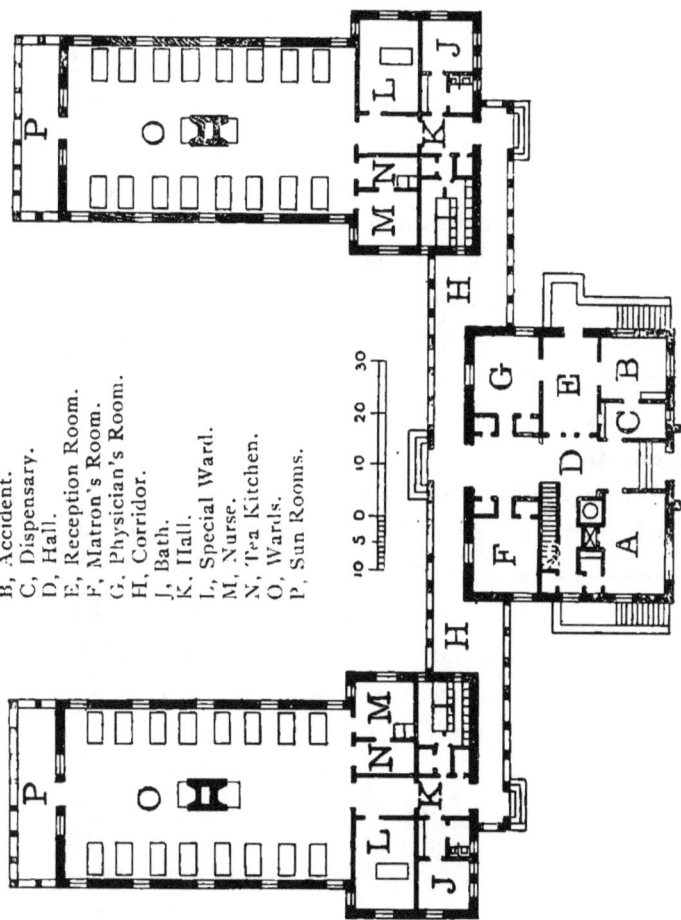

A, Dining Room.
B, Accident.
C, Dispensary.
D, Hall.
E, Reception Room.
F, Matron's Room.
G, Physician's Room.
H, Corridor.
J, Bath.
K. Hall.
L, Special Ward.
M, Nurse.
N, Tea Kitchen.
O, Wards.
P, Sun Rooms.

PLATE 70, PAGE 145.

Plan of the Cambridge Hospital, Cambridge, Mass.

There is no method of heating which can be taken care of by simply cleaning out the boiler tubes occasionally and it should be the duty of the engineer having the mechanical apparatus in charge to periodically have all these coils thoroughly cleaned; much of this filth could be prevented from entering coils and wards by thin cotton screens to be renewed from time to time, which should be placed in the openings where fresh air is admitted.

The location of the registers for admitting fresh air to the wards should be between the beds and near the heads of the beds; in this way a gentle draft is produced so close to the wall that no stagnatiod of air can be produced at this point.

In a small pamphlet entitled a "Contribution to the Study of Ventilation," by Edward Cowles, M. D., he gives observations which were made in the one story surgical ward, of which a plan is given in plate 18, page 37, of this work, and it was found that the "Force of the extracting power exercised by the ventilating chamber is not immediately felt by the air currents below, until within a few feet of the outlets in the ceiling. Above the height of twelve to fourteen feet, the inflowing currents of air seem to have quite lost their initial velocity, and the movement is, as a rule, thence directly upward till deflected towards the central openings by the ceiling." *   *   *   *   *   *   *   *

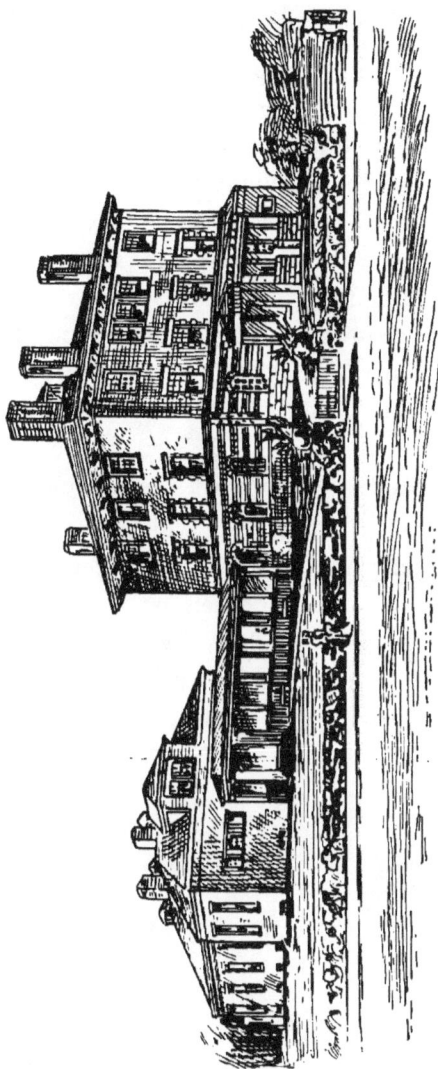

PLATE 71, PAGE 147.

Perspective of the Cambridge Hospital, Cambridge, Mass.

"There certainly appears to be no advantage gained by having the upper six feet of air space, which would not be more than offset by that gained from the more frequent change of the whole atmosphere of the room that the same volume of air supply would give with the lessened air space. It would seem, therefore, that in this case the ceiling might as well be lowered to a height of at most fourteen feet, thus permitting the foul air to escape sooner from the room, instead of occupying the waste space now existing above that level, at the risk of falling again and remixing with the air below."

The height of the ward in which these observations were made, from the floor to the center of the curved ceiling, is twenty feet, or an average of about eighteen and one-half feet; each bed has a clear space of about eighty and one-half square feet , and an air space of about one thousand, four hundred and eighty-nine feet.

CHAPTER XII.

VENTILATION.

In commencing the construction of a hospital, it is invariably the case that every one connected with it announces that the plumbing and sanitary appliances will be of a superior order, "containing all the latest improvements," and that special attention will be paid to heating and ventilation.

The Trustees after "looking up" the matter, individually meet to examine the plans, only to find that each has decided ideas of his own on the subject, and that of all the "systems" none fully agree. Then, with all due regard for every one's opinion, and after modifying his ideas to suit those of the most influential members, some sanitary engineer gets the "job."

The result is, a system of pipes, coils, receivers, dampers, etc., which are bewildering to look at, but which seldom heat well, and it might also be said never ventilate.

After the hospital has been turned over to the manager, he and the physician in charge (whose pet theory has probably been upset) unite in condemning the arrangements, and after a series of expensive experiments, a combination that never has been seen before, and that no one wishes to see again, is arranged, that will produce a current of air in one direction one day, in the opposite direction the second and remain inoperative the third.

But the hospital is new and patients go there because it has been so highly spoken of; the wards are bright and sunny; most of the patients get well, and the "system" of ventilation falls into a state of "inocuous desuetude".

That such a state of affairs exists, is due largely no doubt, to the popular delusion that a flue in a wall with a register at the bottom and an opening above the roof is going to cause a continuous draught of air up it and out at the top.

There are times, no doubt, when this is the case to a limited extent. When the air in the ward is much warmer than the outside air the difference in temperature will cause an upward current, but which will be decreas.d by friction and more or less counteracted by opposing currents of cold air at the top.

But the draught will not usually be permanent, and cold air may frequently be blown into the ward.

In the summer, when the temperature inside the ward may be only a few degrees cooler than the air outside, there is usually no current discoverable.

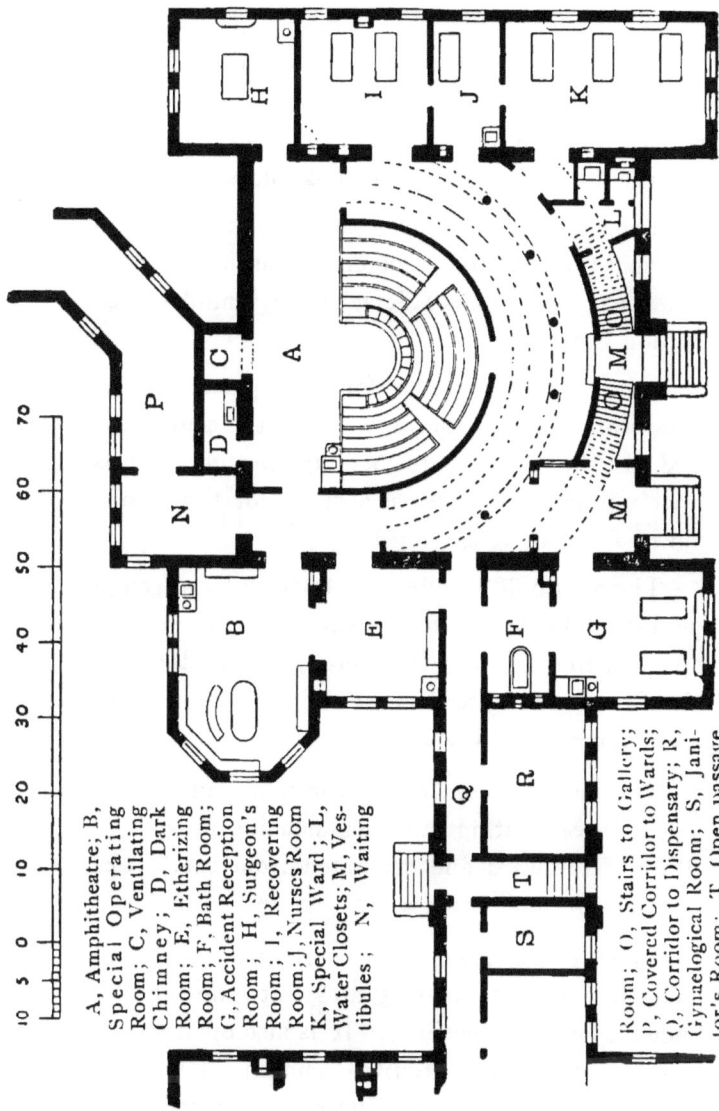

A, Amphitheatre; B, Special Operating Room; C, Ventilating Chimney; D, Dark Room; E, Etherizing Room; F, Bath Room; G, Accident Reception Room; H, Surgeon's Room; I, Recovering Room; J, Nurses Room; K, Special Ward; L, Water Closets; M, Vestibules; N, Waiting Room; O, Stairs to Gallery; P, Covered Corridor to Wards; Q, Corridor to Dispensary; R, Gynaelogical Room; S, Janitor's Room; T, Open passage between Amphitheatre and Dispensary.

PLATE 72, PAGE 151.

Plan of Amphitheatre, Johns Hopkins Hospital, Baltimore.

The only way that such a flue, or series of flues can ventilate is to apply to them a strong continuous artificial heat, either by steam coil, stove, gas jet, or the juxtaposition of a heated smoke flue.

The latter is the surest method, although it is not usual that such an arrangement can be economically made in a series of detached or semi-detached pavilion wards.

In arranging three or four fire places back to back in one stack of mason work, in the centre of a ward, as shown in Plates 6, 9, 35, 51 and 59, it is entirely practicable to leave a space for the admission of fresh air to the ward.

The air would be admitted to the ward through openings four or five feet above the floor, and the supply of fresh air should be brought from the outside of the building to the stack in a galvanized iron pipe below the ward floor.

The smoke flues would also cause an upward current in ventilating flues commencing nearer the ceiling of the ward and separated from the smoke flues only by their brick walls.

This arrangement will work tolerably well when there are fires in the grates, but cannot be depended upon at other times. It is also open to the objection that only a limited amount of warmed fresh air can be admitted owing to the limited heat-

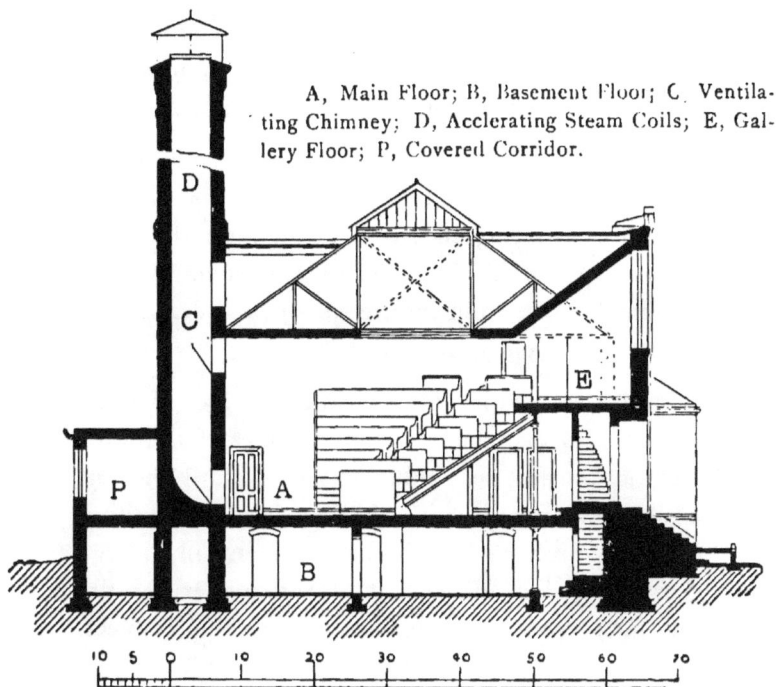

A, Main Floor; B, Basement Floor; C, Ventilating Chimney; D, Accelerating Steam Coils; E, Gallery Floor; P, Covered Corridor.

PLATE 73, PAGE 153.
Section of Amphitheatre Johns Hopkins Hospital, Baltimore.

ing surface of the backs of the fire places, and if used should be re-enforced by other arrangements.

This is the only way that heat can be economically applied, without mechanical force, for the purpose of ventilating in pavilions, because the units of heat that would be otherwise wasted, pass up the smoke flues and induce an upward draught in the adjacent ventilating flues.

The cost of providing heat at the foot of ventilating flues arranged around the sides of the walls for the purpose of inducing an upward current would be more expensive than the cost of running a fan to produce by suction a current in a large ventilating shaft into which all the smaller ventilating ducts from the various pavilions could connect.

Probably no means of ventilating has been condemned so extensively, and with such unanimity of opinion, as the fan. Frequently because it was not understood—more frequently because the cost of running it has, until recently, been an unknown quantity.

It is only recently that electricity, as a mercantile commodity, has been offered to the public, like gas and water, for a stated price per month. With the advent of electricity the most serious objections to the use of the fan as a means of ventilation have been removed. Heretofore, it has been necessary,

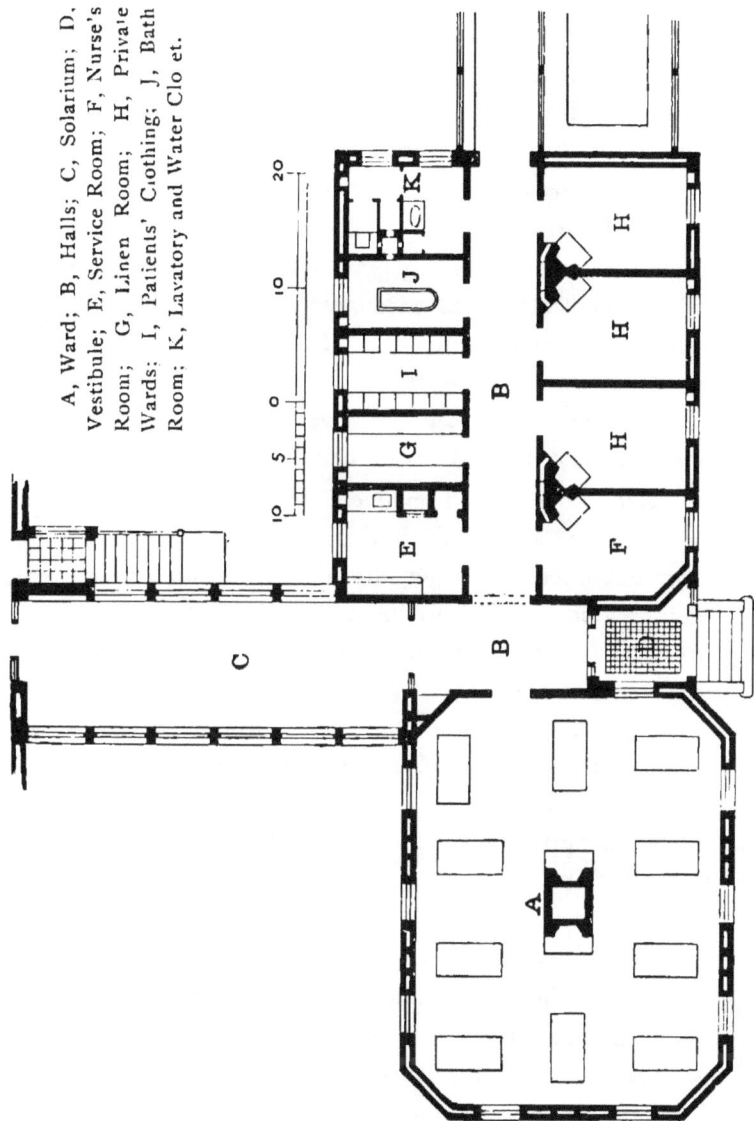

A, Ward; B, Halls; C, Solarium; D. Vestibule; E, Service Room; F, Nurse's Room; G, Linen Room; H, Private Wards: I, Patients' Clothing; J, Bath Room; K, Lavatory and Water Clo et.

PLATE 74 PAGE 155.

to obtain efficient service to keep up the steam the year around and maintain a more or less expensive engine for running the fan.

This necessitated a skilled engineer and fireman, and occasioned large bills for repairs. Now any local electric lighting plant can furnish power by means of its wires to an electric motor which will revolve the fan with little attention. The sizes of these motors vary from one-half H. P. up; it being seldom that one of   greater than two and one-half H. P. capacity is needed.

Such a motor will revolve a fan 42″ in diameter at a speed of 250 revolutions per minute, and move 7,079 cubic feet of air per minute. This speed may be increased, and the air moved will be increased in proportion. The care of the fan and motor can be easily undertaken by any one possessing ordinary intelligence as it can be made "self oiling," and the amount of time needed to keep it in order should not exceed one hour per week. Of course, in hospitals warranting such an expenditure, or in institutions of any size where the funds for construction will admit, an entire electric plant may be put in for furnishing this power and light.

But if the institution is a small one and the power is bought at a stated price per horse power per month, it can be utilized, besides running the fan, for turning the washing machines in the laundry

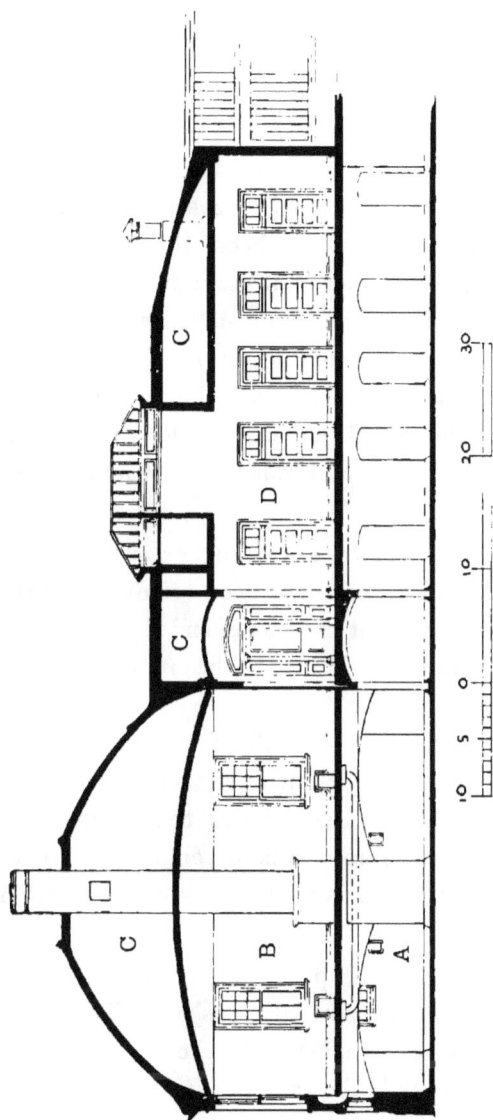

PLATE 75. PAGE 157.

Section of Ward in the Mary Hitchcock Memorial Hospital. Hanover, N. H.
A, Basement; B, Ward; C, Air space between Ceiling and Roof; D, Hall.

and running the sewing machines in the sewing room.

This of course would be accomplished at a slight expense by means of a counter shaft and pulleys. Heretofore it has been the custom also to decry the fan as a means of ventilation because it was claimed that sufficient appropriations for keeping it constantly running and for repairs would be cut down by parsimonious boards having charge of public institutions and that when it fell into disuse from neglect of lack of funds there would be no ventilation.

While this was to a certain extent true, all danger of this kind has now been removed and the subtile electric current can be depended on to do its work night and day with all the regularity that could be desired.

No one who has used the fan for ventilating and felt the powerful current of air being constantly thrown out by it, can have failed to be impressed with its value as a sure means of removing foul air no matter what the conditions of wind or weather might be.

The fan should be placed at or very near the outlet of the main ventilating shaft. The shaft might be of brick or metal. It should have, if of metal, tight joints. The arrangement of it should be simple in its branches.

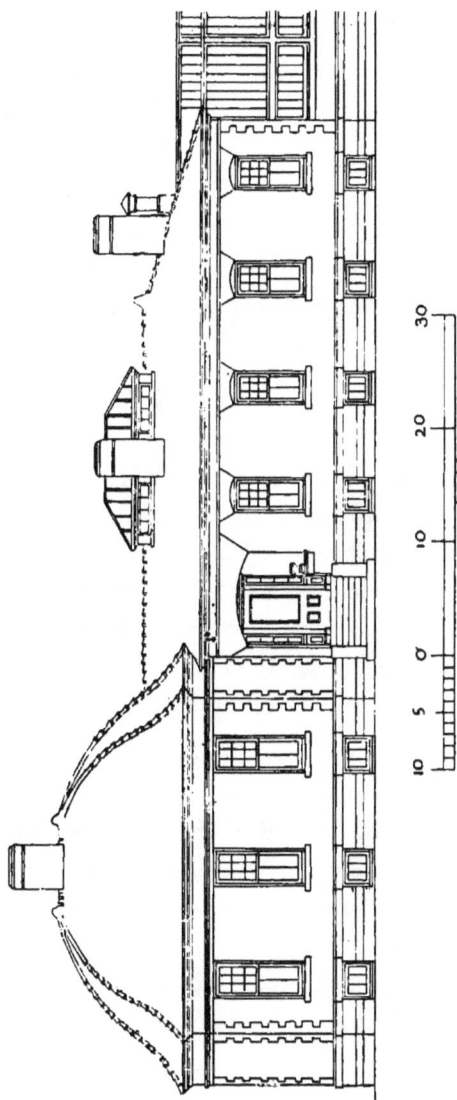

PLATE 76, PAGE 159.

Elevation of Ward in the Mary Hitchcock Memorial Hospital, Hanover, N. H.

Every pipe from every ventilating flue should increase in proportion to its area as other pipes are connected to it, due allowance being made for friction. The pipes should be ample in size, so that the fan need not be run at the maximum speed, or cause too swift a current, creating a draft over the patients. At the same time a positive and sure draft should be at all times maintained.

The ventilation ducts should be formed of some non-absorbent substance—never of wood, and of galvanized iron in preference to brick. The surfaces of brick walls and the rough mortar joints catch innumerable particles of dust and also cause a great loss of effective energy owing to the increased friction. Galvanized iron, while not as smooth as tin, can be obtained in large sheets thereby lessening the number of joints.

As tin is a poorer conductor than galvanized iron, tin pipes should be used for supplying heat, as little is lost by radiation before entering the wards, but after the vitiated air has left the wards in the ventilating ducts the use of the galvanized iron is preferable to tin, as by radiation, more of the otherwise wasted heat units are saved within the walls of the building.

As heated air naturally rises, of course there is more friction in pulling it down than up, but other-

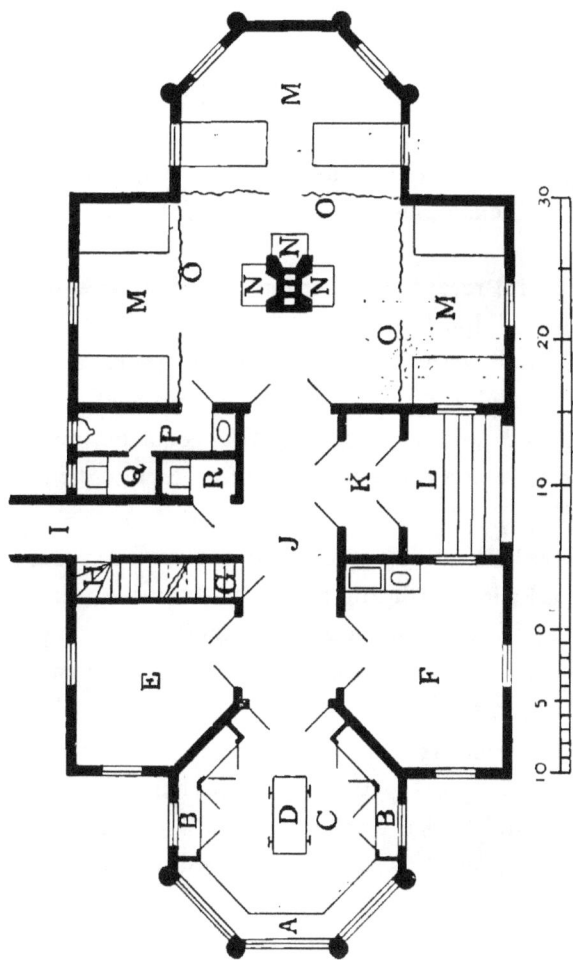

PLATE 77, PAGE 161.

Floor Plan of an Inexpensive Operating and Emergency Annex.

wise it is immaterial whether the ventilating ducts lead to a main ventilating shaft in the basement or in the attic.

Local conditions must govern this, and many successful examples of both locations might be quoted.

Several recent and serious hospital fires have shown the importance of using every precaution to prevent their occurrence. Too great care cannot be taken in buildings that are not fire proofed, and it is the duty of those having the construction in charge to lessen the possibility of an incipient fire's spreading.

For this reason (especially if the roof is framed of wood) the ventilating shaft should not be placed in the attic, where in case of fire the forced draught would prove a sure means of almost instantly spreading it beyond control.

The certainty that such a fire might prove disastrous to helpless patients, and perhaps (from exposure) fatal to many others, should overweigh the slight additional cost of ventilating downwards.

The fan being located in the basement instead of the attic is more directly under the eye of the engineer, and when a large fan is used there is less vibration, although there is never very much. The foul air should never be discharged out of a basement opening to the outside air, as it might be

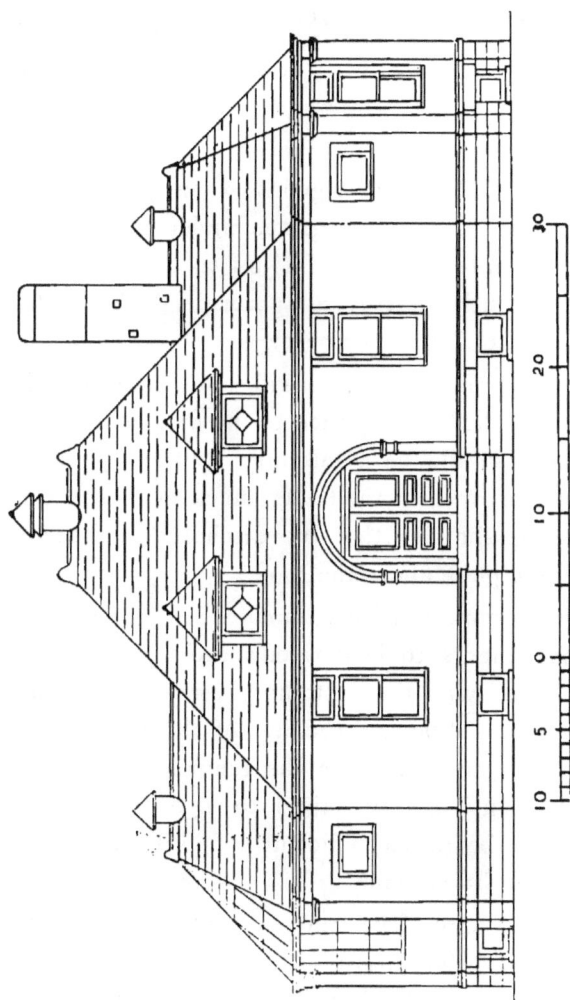

PLATE 78, PAGE 163.

Front Elevation of an Inexpensive Operating and Emergency Annex.

drawn back in a diluted condition through the openings supplying fresh air to be re-breathed by the patients.

It would be advantageous when possible to discharge it through the grates under the boilers, where the intense heat of 1,500 to 2,000 degrees would instantly destroy all germs, and thence by the smoke stack into the outer air.

Usually the grate area will not admit of this forced draught being applied to the fires, and therefore there should be a masonry ventilating stack built side of or near the boiler stack to receive the ventilating shaft so that the discharge of foul air would take place above the roof.

Or, without great additional expense (considering the benefits to be derived thereby,) the boiler might be set with the customary air space in the setting enlarged beyond the area of the main ventilating shaft, so that while the temperature of the foul air would not be raised to 2,500 degrees it would still be raised sufficiently to kill all germs which it might contain.

Another way of heating the air from the ventilating shaft before passing it out would be to exhaust it into the "uptake," viz., the point in the vertical smoke stack just above its connection with the horizontal smoke pipe. Either of these methods must be adopted only upon the advice of some

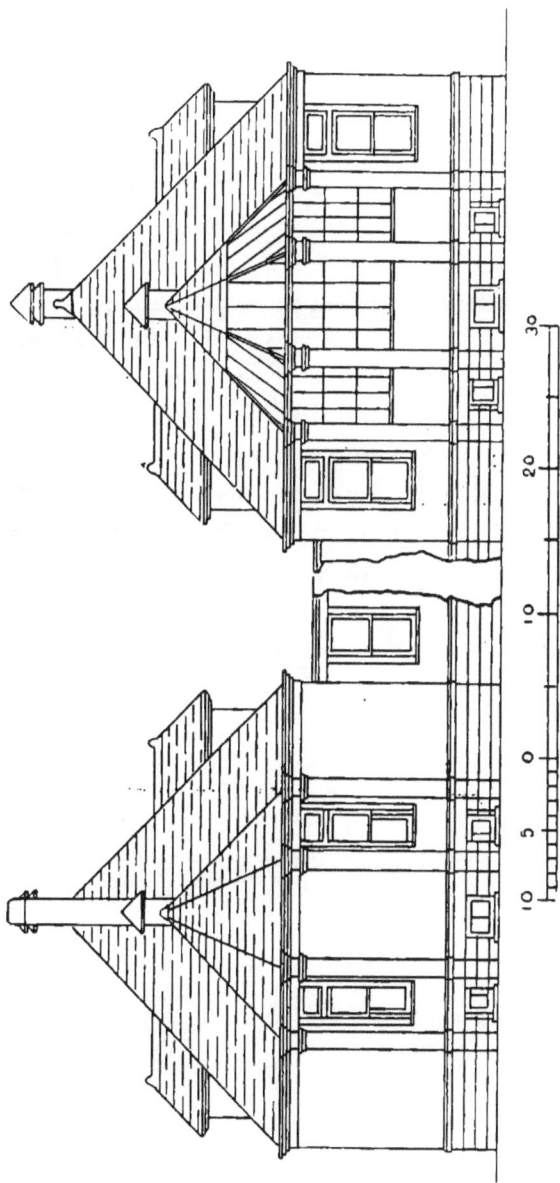

Elevation of each end of an Inexpensive Operating and Emergency Annex.

expert engineer.   The temperature in the "uptake"
usually is from 450 to 700 degrees, according to the
method of setting the boiler, and while such a
temperature might not destroy *all* the germs, it certainly would some, and the general results would be
decidedly beneficial.

A very simple, and on the whole perhaps better,
arrangement is to build a wrought or cast iron
smoke stack in the centre of the masonry ventilating stack : when this is done it should not be forgotten that in the ring, formed by the concentric
surfaces of the exterior of the smoke stack and the
interior of the ventilating *stack*, there would be
much more friction than in the main ventilating
shaft, and that consequently the area of the ring
should be considerably increased.

CHAPTER XIII

In her Notes on Hospitals, Florence Nightingale says: "It was found in a certain Parisian hospital in which the ventilating arrangements were deficient, that pyæmia and hospital gangrene had appeared among the patients. These diseases are said to have disappeared at the introduction of ventilating arrangements, whereby 2,500 cubic feet of air per bed per hour were suppled to the wards.

Notwithstanding this large quantity, however, the ward atmosphere was found not to be sufficiently pure. In other wards the quantity of air was increased to as much as four to five thousand cubic feet per bed per hour."

One of the chief objections which has been raised in the past against artificial ventilation, has been the statement made by certain writers on hospitals, that by it fresh air is seldom introduced to a ward and that foul air is seldom extracted. While this may have been true of the mechanical arrangements of an earlier date, such is not the case now; yet these books, many of them published nearly

thirty years ago, are still referred to as standard
works on hospital construction and arrangement.

We find writers advocating radiant heat as
being the *only* proper way of heating hospitals.
Aside from the dangers from fire in buildings constructed to obtain the greatest accommodation at the
least cost, we have the attendant dust and dirt, and
the large amount of service required to properly
take care of a large number of fires in different
rooms.

We are, of course, by this method assured of a
continuous ventilation—when there is a fire. When
there is no fire the ventilation is slight, and we are
told we can rely on the "magnificent supply of windows" for fresh air. We are told that to supply
each bed from 2,600 to 5,000 cubic feet of fresh air
per hour, at a uniform temperature, by the mechanical means is not satisfactory.

The criticism is also made that it is not in ac-
cordance with nature's method of providing fresh
air ; or, "she affords air both to the sick and healthy
of varying temperature, at each hour of the day and
night and season—always apportioning the quantity of moisture to the temperature, providing continuous free movement everywhere, and warming
not by warm water in iron pipes, but by radiated
heat. We all know how necessary the variations of
weather, temperature and season are for maintaining

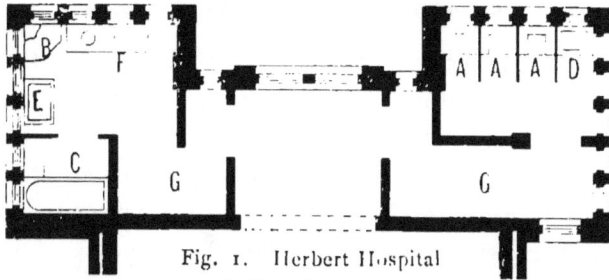

Fig. 1. Herbert Hospital
Toilet Pavilions,
Woolwich.

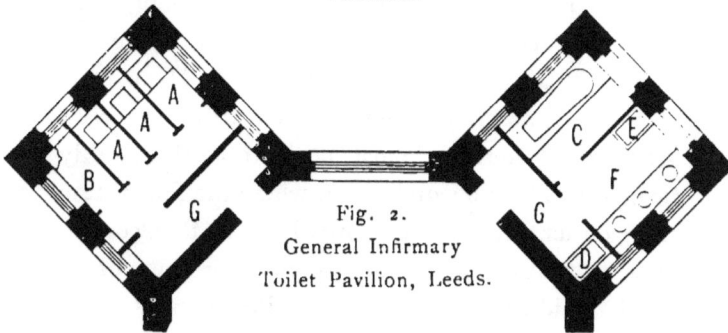

Fig. 2.
General Infirmary
Toilet Pavilion, Leeds.

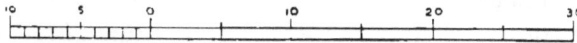

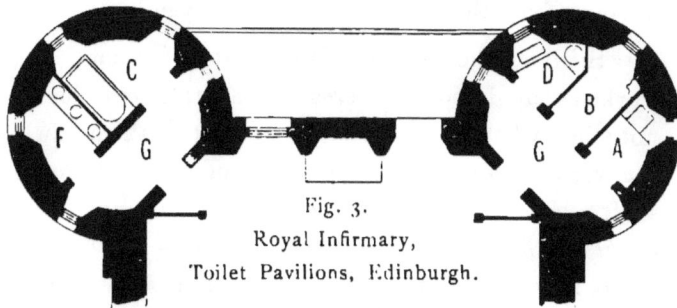

Fig. 3.
Royal Infirmary,
Toilet Pavilions, Edinburgh.

PLATE 80. PAGE 169.

A, Water Closets; B, Urinals; C, Bath Rooms; D, Slop Sinks;
E, Sinks; F, Wash Stands; G, Lobbies.

health in healthy people. Have we any right to
assume the natural law is different in sickness? In
looking solely at combined warming and ventilation
to ensure to the sick a certain amount of air at 60
degrees, paid for by contract, are we acting in ac-
cordance with physiological law?"

We are told by other writers that "if the continuous use of a fan is not feasible, it should certainly be driven every morning and evening, say
from 5 A. M. to 8 A.M., and from 7 P. M. to 10 P.
M., for it is at these times that the air of the ward
grows most foul and most needs to be thoroughly
cleaned out." No wonder that where the fan system
has been used in such a criminal way as this, continuous ventilation and properly pure air was not se-
cured. Did the writer of these lines suppose that
patients did not breathe or that carbonic acid was
not produced during the other 18 hours of the 24
during which the fan was not to be driven?

Still another writer states that a fan may be
worked "by hand" for creating a draft, but does
not state whether it is simply to agitate the air in
the ward or to remove it by means of a ventilating
shaft. *If a fan is used it must be used continuously
night and day. Unless this can be done*—and there
is no reason why it cannot be done as surely as
steam can be kept up in a boiler, or as gas can be
produced continuously in pipes, or as water can be

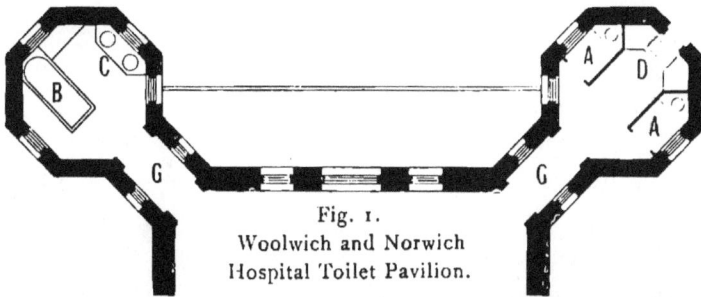

Fig. 1.
Woolwich and Norwich
Hospital Toilet Pavilion.

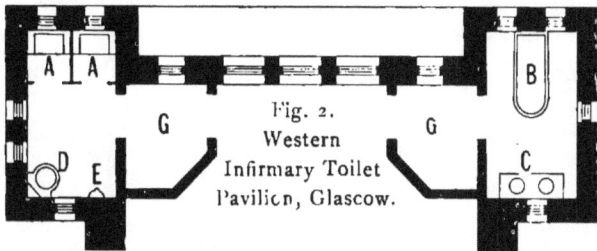

Fig. 2.
Western
Infirmary Toilet
Pavilion, Glascow.

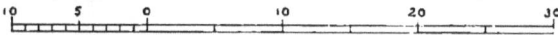

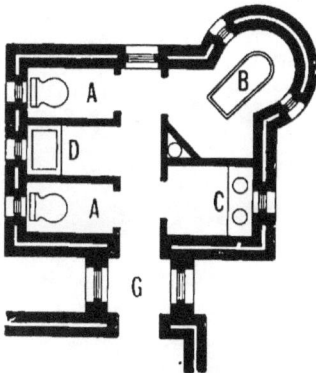

Fig. 3. Royal Infirmary
Toilet Pavilion, Liverpool.

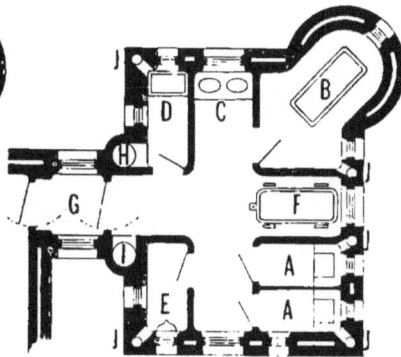

Fig. 4. Suggestion for a Toilet
Pavilion.

PLATE 81, PAGE 171.

A, Water Closets; B, Bath Tubs; C, Wash Stands; D, Sinks;
E, Urinals; F, Movable Bath Tub; G, Lobbies; H, Chute for
Bandages, Rags, etc.; I, Solid Linen Chute; J, Ventilation Flues.

pumped incessantly by corporations, *it should not be
used at all.*

It is not the intention of the writer to decry
natural ventilation ; it is valuable not only in large
wards, but in rooms in hospitals or dwelling houses ;
but when people make such ridiculous assertions as
have been made regarding the use of the fan for
ventilating and heating hospitals, it is time some
one contradicted them and put the matter in a fair
light before those who are interested in these institutions.

END OF VOL. I.

INDEX.

A

B

C

IV.

H

I

J

L

M

P

R